TROPICAL FISHES OF THE GREAT BARRIER REEF

TROPICAL FISHES OF THE GREAT BARRIER REEF

Tom C. Marshall

Government Ichthyologist 1943-1962
Department of Harbours and Marine, Queensland

Illustrated by George Coates

Angus & Robertson Publishers

Angus & Robertson Publishers
London . Sydney . Melbourne . Singapore . Manila

First published by Angus & Robertson Publishers, Australia, 1966
This revised edition 1982

National Library of Australia
Cataloguing-in-publication data.

Marshall, Thomas Claude, 1897-1976.
Tropical fishes of the Great Barrier Reef.

Index
ISBN 0 207 14157 6

1. Fishes—Great Barrier Reef. 2. Fishes—Tropics.
1. Coates, George, illus. II. Title. (Series)

597.09943

Type 10pt Baskerville by Setrite Typesetters, Hong Kong
Printed in Hong Kong

Contents

Metric Conversion Guide

In this updated edition metric equivalents have replaced the imperial measurements of the first edition. The metric tables below are *approximate* guides.

However, in the interests of fluency an exception has been made with "fathoms" and this old nautical measurement of depth has been retained throughout. 1 fathom = 2 yards = 1.828 metres.

Length

centimetres to inches

cm		in	cm		in
0.5	=	¼″	45	=	18″
1	=	½″	48	=	19″
2	=	¾″	50	=	20″
2.5	=	1″	53	=	21″
5	=	2″	56	=	22″
7.5	=	3″	58	=	23″
10	=	4″	60	=	24″
13	=	5″	63	=	25″
15	=	6″	65	=	26″
18	=	7″	68	=	27″
20	=	8″	71	=	28″
23	=	9″	73	=	29″
25	=	10″	76	=	30″
28	=	11″	79	=	31″
30	=	12″	80	=	32″
33	=	13″	83	=	33″
35	=	14″	86	=	34″
38	=	15″	88	=	35″
40	=	16″	90	=	36″
43	=	17″			

cm/metres to feet

cm/metres	feet
30	1
60	2
90	3
1.25	4
1.5	5
1.75	6
2	7
2.25	8
2.75	9
3	10
3.25	11
3.5	12
4	13
4.25	14
4.5	15
4.75	16
5	17
5.5	18
5.75	19
6	20
15	50
30	100

Weight

pounds to kilograms

lb		kg
1	=	.45
2	=	.90
3	=	1.35
4	=	1.80
5	=	2.25
10	=	4.50
15	=	7
20	=	9
25	=	11.50
30	=	13.50
35	=	15.90
40	=	18
45	=	20.50
50	=	22
100	=	45
500	=	226
1000	=	452
2000	=	904
1 ton (2240 lbs)	=	1016

Editor's Note

Numerous names have been changed since the original publication of *Tropical Fishes of the Great Barrier Reef* in 1966. Helen Larson of the Department of Ichthyology at the Australian Museum in Sydney has brought the scientific names into line with those currently in use today, and in a few cases has changed the common names.

Illustrations are provided for three species which were described in previous editions but not pictured. These are the Bronze Whaler and the Butterfly Gurnard (courtesy of the Australasian Marine Photographic Index) and the Thorntailed Pipe-fish (courtesy of Helen Larson).

Acknowledgements

ILLUSTRATIONS *in this book, other than those by the author, are taken mainly from the water-colour drawings of George Coates, of Townsville, North Queensland, and also from the work of other artists and from various ichthyological publications and papers, according to plate references given here. The use of these illustrations is gratefully acknowledged by the author.*

Amer. Mus. Journ. — Pl. 73 (jaws of white shark)
Ann. Mag. Nat. Hist. (Lond.). — Pls 29:5; 134:5
Atlas Ichthyology, Bleeker — Pls 90:6; 91:3; 93:3; 95:1, 35; 97:5, 7; 102:3; 103:3; 104:1; 150:3; 106:5, 6; 107:5; 108:1, 2; 111:1, 2; 112:2; 114:1, 2, 4, 5, 7; 115:2, 4; 116:3, 4, 5, 6; 117:4, 5; 118:1, 2; 119:2; 129:5; 132:5; 133:1; 134:7; 136:7
Austr. Mus. Mag. — Pl. 123:5
Austr. Zool. — Pls 76:5; 78:4; 89:1, 3; 98:6; 102:4; 103:4; 124:1; 127:4; 132:1
Biol. Res. Endeavour. — Pls 86:2; 87:1; 98:3; 101:3; 105:1; 106:4; 109:5; 110:1; 112:6; 123:7; 129:6, 7, 8; 130:3; 131:2, 3, 5, 6, 7; 132:2, 6; 136:1
Bull. U.S. Fish. Comm. — Pls 97:4; 117:6
Bull. U.S. Nat. Mus. — Pls 77:2, 3, 4; 99:2
Coates, G. — Pls 1-31; 32:1, 2; 33-43; 45:1, 2; 46-47; 48:2, 3; 49-72; 75:1; 76:5; 78:4; 81:1-4; 83:1; 84:5; 85:1; 86:4; 91:6; 92:2, 5, 8
Coleman, N. — Pls 77:2; 131:10
Courier Mail, Qld. — Pl. 108:4
Edib. Fish N.S. Wales, Stead — Pls 90:4; 96:2; 111:3; 114:3
Edib. Fish Crust. N.S. Wales, Ogilby — Pl. 115:2
Fauna Japonica Pisces — Pl. 135:4
Fish. Gt. Brit. and Irel., Day — Pls 103:1; 133:6
Fish. Indo-Austr. Archip., Weber & de Beaufort — Pls 98:4; 99:3; 100:4; 115:6
Fish. Japan, Tanaka — Pl. 135:3
Fish of Austr., T.C. Roughley — Pls 108:6; 134:2

Fish of India, Day — Pls 77:5, 6; 84:1, 3; 87:2; 88:5; 93:5; 95:2, 3; 97:6; 98:5; 100:2; 119:4; 129:2; 134:3

Fish. Sth. Austr., Waite — Pls 74:1; 76:2; 79:6; 87:5, 6; 93:2; 100:5; 130:10

Flat-fishes (Heterosomata), Norman — Pl. 132:3

Gt. Barr. Reef Exped. Brit. Mus (Fishes), Whitley — Pl. 128:3

Hall, Miss M. — Pls 105:3; 107:3; 112:3; 125:5

Hist. Nat. Poiss., Cuv. and Val. — Pl. 117:3

Ichth. Notes, Dept. Harb. and Marine, Qld. — Pl. 92:1

Journ. Coll. Agric. Tokyo, Kishonouye — Pls 120:3; 122:4

Journ. Mus. Godeff — Pls 96:1; 98:1; 99:1; 104:2; 127:2; 129:3; 134:1; 136:4

Journ. Proc. Roy. Soc. N.S. Wales — Pl. 79:3, 4, 5

Know Your Fishes, Dept. Harb. and Marine, Qld. — Pls 105:2; 106:7; 112:3

Marine Dept. Rept. Qld. Tosh — Pls 111:4; 128:2

Marshall, T.C. — Pls 20:2; 32:3; 44:1; 45:2; 48:2; 54:1

Mem. Austr. Mus. — Pls 85:2; 86:1; 98:2; 112:1; 114:6; 117:8; 133:2

Mem. Bern. P. Bishop Mus. Fish. Oceania, Fowler — Pl. 127:1

Mem. Qld. Mus. Pls 48:1; 74:1, 2; 79:1; 85:4; 91:4; 96:3; 104:3; 106:2; 107:6; 108:3, 5; 109:1-4; 110:2-4; 111:6; 115:2, 3, 7; 118:6, 8; 119:3; 121:1, 3: 122:1-3; 124:6; 125:1; 128:4; 130:2, 5, 6; 133:3, 5; 134:4, 8

Müller and Henle, Plagiost — Pl. 87:6

New Zeal. Journ. Sci. Tech. Pl. 93:4

Novara Zool. Fische — Pl. 117:1

Olsen, F. — Pls 131:1; 135:2

Olsen, F. (after McCulloch and McNeill) — Pl. 124:2

Olsen, F. (after McCulloch and Whitley) — Pls 131:1; 133:4

Olsen, F. (after Müller and Henle) — Pl. 78:1

Olsen, F. (after Norman) — Pl. 85:5

Olsen, F. (after Whitley) — Pls 75:4; 76:3, 4; 77:3; 78:2, 3; 79:2; 80:3; 86:3; 89:4; 90:5; 95:4; 124:8; 129:9; 130:9

Oceanic Ichthy., Goode and Bean — Pl. 125:2

Phil. Journ. Sci. — Pls 106:3; 118:3, 5, 7; 119:1; 124:5

Proc. Linn. Soc. N.S. Wales — Pls 76:1; 77:1; 85:6, 7; 99:6; 119:5; 123:8; 136:6

Proc. U.S. Nat. Mus. — Pls 44:2; 118:7; 131:8

Proc. Zool. Soc. Lond. — Pls 80:1, 2; 113:1, 2; 124:10; 134:6

Prod. Zool. Vict., McCoy — Pls 82:1, 2; 91:1; 101:2; 102:1

Public Relations Bur. Qld. — Pl. 84:2

Rec. Austr. Mus — Pls 78:5; 80:4, 5; 89:5, 6; 90:1-3; 91:5; 92:7; 93:1; 96:4; 99:5; 100:3; 101:1, 4; 103:2; 105:5; 106:1; 107:4; 110:5, 6; 112:4, 7; 115:1; 116:7; 117:2, 7; 123:1-4, 9, 10; 124:3, 5, 7; 126:1, 3; 130:7, 8; 132:7; 135:5

Rec. Sth. Austr. Mus. — Pls 82:3; 136:5

Rec. West. Austr. Mus. Pls 99:4; 135:8

Rept. Voy. Challenger, Günther — Pls 102:2; 132:4

Roughley, T.C. — Pl. 73

Sea Fish Sthrn. Afr., Smith — Pls 125:3; 126:2; 127:3; 128:1

Sitzb. Akad. Wiss. Wien — Pls 111:4; 112:5; 116:1; 124:12; 129:1; 135:1

The Fish Austr. (Sharks), Whitley — Pls 76:3; 77:4; 84:4; 86:5; 87:4; 88:1, 5

Trans. New Zeal. Inst. — Pl. 121:1

Trans. Roy. Soc. Sth. Austr. — Pls 118:4; 125:4

Voy. Astrolabe (Zool.) — Pl. 131:4

Voy. Erebus and Terror — Pls 89:2; 91:2; 92:3, 4; 124:11; 131:9; 134:9; 135:6

Fishes of the Reef Waters

INTRODUCTION BY GILBERT WHITLEY

Everybody who visits the Great Barrier Reef, even if not a naturalist or angler, is impressed by the fishes there. The variety of species, beauty of colouring, strangeness of shape, to say nothing of the delicacy of taste of some of the commercial kinds, all command attention and admiration. Indeed it would be difficult, anywhere in the world, to equal the variety of the fishes found in the State of Queensland.

Only the more important kinds of fishes can be dealt with in an introduction of this kind and the freshwater ones can at once be dismissed since only small inconspicuous forms inhabit the very few rills and rivers of some of the larger islands off Queensland's tropical mainland.

The saltwater species are evidently derived from an enormous area of warm seas stretching from the Red Sea to eastern Africa and then eastward in a wide belt through India and the Indo-Pacific archipelagos to Hawaii and the south-eastern Pacific Islands; from this area our northern Australian marine fauna comes down to below the Tropic of Capricorn on both western and eastern sides. There are, however, a number of species which are known only from Queensland but they probably extend at least around our northern shorelines and may be found over a wider range when the seas are better explored.

The visitor to the Great Barrier Reef experiences some of the thrill of the explorer for, though no hardships of thirst or cold need be suffered, no tremendous peaks need be scaled, there is always the likelihood of finding some new fish in the more modest scope of rock pool or coral reef in the beautiful garden-like grottoes, off the clean sandy beaches, or in the still largely uncharted open waters which lie beyond.

I propose to deal with the Barrier Reef fishes according to the regions in which they are found, for fishes do not swim indiscriminately everywhere but have their favoured retreats, their laneways for migration, certain water-temperatures and depths at which they are at their best, and so on. These limits, though perceptible to us, are not absolutely hard and fast, but are elastic at some places and times so that the fishes are influenced by the presence or absence of food or enemies, the rise and fall of tides, the scour of currents, the fury of storms, and the pulse of the mighty ocean itself.

The mud-loving fishes of a mangrove swamp are so different from those of the clear waters of a coral reef or a sandflat that they might almost be inhabitants of another continent, yet such is the complex evolution and structure of many of the islets of the Great Barrier Reef that members of several faunal regions may be found around a single given island. Mud-skippers may hop over the mud of the mangroves yet on an adjacent shingle bank there will be entirely different fishes.

The coral-seekers are legion and their colours and camouflage of never-ending interest, yet almost as varied are the food-fishes to be hooked or netted from a nearby sandy beach or sought in the more open waters between the Barrier Reef and the mainland. Also the ceaseless pounding of giant breakers on the Outer Barrier reminds us constantly of the big game-fishes in the open ocean and the weird, unknown curiosities of the deepest water beyond which has so far not been explored by human beings.

Fishes show, in their structure, some clues to their haunts and habits. Some have long snouts and delicate bristle teeth for insertion into the interstices of coral in search of food. The Parrot-fishes have beak-like teeth and grinders in their throats for crunching chunks of coral to find their nourishment. Others like Eels and Pike have long and pointed fangs.

Gaudy colours which would make a fish conspicuous in colder seas are really helpful camouflage in the kaleidoscopic variety of colour and form in a living coral reef. None the less, the sand-haunting fishes (Flatheads and Flounders for example) closely imitate their background in colour; the species of the mangrove mud are dull in hues, and every fish is on guard against attack from its enemies as in nature universally. In shallow reef-pools some fishes are closely associated with certain "hosts" (sea-anemones, sea-weeds and particular types of coral) which shelter them almost exclusively.

So on the Great Barrier Reef anyone interested in fishes will find thousands of those fascinating but little understood creatures, from tiny Gobies whose whole life may be spent in a little coral garden and which lay their eggs and guard them with their tiny lives, to the giant sharks and rays which cruise where they will and the swift, streamlined game-fishes which, for all we know, may speed round half the world in the course of their careers.

OCEAN SURFACE FISHES

For giant fishes and those sought by big game anglers it is necessary to brave the open sea. The waters outside the Great Barrier Reef are little explored from the fisherman's point of view and are often too rough to suit the average holiday maker, whilst most of our shipping uses the route between reef and mainland.

Reports of giant ocean Sunfishes and Thresher Sharks come from open water off the south Queensland coast, but these creatures are more inhabitants of the south which are wandering beyond their usual range. The Sunfish is a huge disk-shaped fish with a long dorsal fin above and a similar anal fin below and may measure 4 metres between the tips of those fins, or 3 metres from its snout to the end of its stumpy tail. It lies on the surface on one side, at times, as if basking in the sun; or couples of them may swim along with the dorsal fin partly out of water. They are quite harmless and feed on the plankton, that floating population of tiny sea-animals and plants found in almost all seas. Sunfishes produce millions of eggs and very young ones are globular, spiny fishlets which are eaten in great numbers by other fishes. This is just as well since, if it were not so, theoretically at least, the sea would eventually become choked by giant sunfishes.

The Thresher Shark is distinguished from all other sharks by having its tail longer than the combined head and body. It grows to over 5 metres in length and feeds on small school fishes which it herds into a mass by repeated blows with its tail on the water or it may maim a fish with its threshing tail before eating it. Stories are told of Thresher Sharks attacking whales but the shark's small mouth and teeth are not adapted for feeding on whale flesh and reports of attacks by threshers may be based on the savage attacks on whales by their smaller but more savage relatives, the Killer Whales.

Sometimes great Devil Rays are seen cruising along the surface of the open sea. They are quite harmless in spite of their size (at least 4 metres wide) and feed on plankton which they steer into their wide mouths by means of flippers on each side, filtering their food through their gill-arches and gill-slits. Devil Rays and Spotted Eagle Rays may be seen (and heard) leaping right out of the water and falling back with a tremendous splash and bang. They probably do this from sheer joy of living, to have a look around, to get rid of parasites, or to assist in the birth of their young.

By trolling lures and the use of big game tackle, a variety of sporting fishes can be obtained in open waters, especially over sunken reefs. The Pike, known as Dingo Fish in Queensland, grabs the bait in his great fangs and is soon secured on a double hook (one hook threaded through the eye of another). These fish are very savage and even when lying on deck are likely to snap at anyone's hand or leg as they see it flash near; the usual practice is to hammer them on the head to quieten them. They are closely allied to the Barracuda of the West Indies which has often attacked man. No fatality due to Pike has been reported from Australia, however.

Various types of Tunnies may be seen disporting themselves. These fish are large members of the Mackerel family and have a number of finlets above and below the hinder part of the body; some of the fins fit into streamlined grooves and the tail fin is widely forked, all marks of strong swimmers. The commonest are the Northern Bluefin Tuna which often leaps right out of the water during calm weather and the smaller Mackerel Tuna (*Euthynnus*) which just arches its back above the surface. Other kinds of Barrier Reef Tunas are the Yellowfin, the Dogtooth Tuna, the Large-scaled Tunny and the Leaping Bonito. Captain Cook saw several "Albecores", as he called the Tunnies.

A close relative of the Tunnies is the Kingfish or Spanish Mackerel, one of the most important fishes in the economy of Queensland, over 100,000 kilograms of it being marketed in a normal year. There are several species, none of them the same as the fishes called "Kingfish" in other Australian States, but all have the tunny-like finlets, but longer, compressed bodies usually adorned with wavy bars or dull spots. The young are known as Doggies. The eggs and young are found in North Queensland, several million eggs being produced by a single female; she then migrates south to feed on fish, squid and crustaceans in preparation for the next breeding season. The natives of Torres Strait made string figures, dancing masks and drawings to depict this common Kingfish.

A large relative of the Queensland Kingfish is the Wahoo. The big game record for this species was a Hawaiian fish, 2.1 metres long and weighing 61.6 kilograms, but Australian examples so far caught have been less than 43 kilograms in northern New South Wales, Queensland and north-western Australia. Even larger are the several sorts of Swordfishes or Marlins so much sought by game anglers. There is, unfortunately, very little accurate scientific information on the species of Marlins found in Queensland but there is no doubt that more than one kind occurs there and some good catches have been reported; one Townsville fish, indeed, had *two* swords for good measure. In spite of reports that Broadbill Swordfish "breed in millions" in the Gulf of Carpentaria, few fishermen have taken one in Queensland and most of the Barrier Reef Swordfish are evidently Marlins. These have the "sword" round in cross-section, not blade-shaped as in the Broadbill. The Black Marlin Swordfish farther south grows to over 4.5 metres and 556 kilograms and the Striped Marlin is smaller; there are reports of Blue and White Marlins as well but as few specimens have been submitted to scientists for identification it is at present impossible to say for sure what species occur in Queensland. The Sailfish is more readily recognized since it has a very high sail-like dorsal fin and a long bayonet-like sword.

When a big game fisherman lands a Swordfish he usually has his photograph taken with it and forthwith loses interest in the fish apart from perhaps keeping the sword. It would be of value to future tourists, fishermen and naturalists if they would have their specimens examined by museum specialists and further knowledge would be revealed by examining the stomach contents, roe, and such-like, and tabulating the information alongside the locality and date of capture of the fish.

Other open water fishes likely to be caught on trolling lines are these: The Sergeant Fish, very shark-like in the water, yet a true fish with hard, dark brown head and brown bands along the body. It is said to exceed 65 kilograms in weight. The Turrum, a large slightly spotted kind of Trevally with the characteristic row of "scutes" or sharp-ridged scales along each side of the butt of the tail. It grows to about 3 metres and 41 kilograms in weight. The Runner, an elongate kingfish-like species with blue and yellow bands reaching from snout to tail; it favours submerged reefs in the open sea and exceeds 5 kilograms in weight. Several kinds of Whitefish, Queenfish, or Leather-skins which are related to the trevallies but have no scutes and the scales are flake-like and buried in a leathery skin. On the sides this skin is very white and makes an excellent and enduring lure for catching other fishes, standing up to bite after bite and proving an irresistible attraction. Some Queenfish grow to 11 kilograms. The Dolphin-fish, famed for its glorious colours, blue, yellow and grey or brown but changing from one to the other before its captor's eyes. It is a true fish with a many-rayed dorsal fin all along the back, not a mammal like the small whales known as Dolphins. Dolphin-fish, up to about 2 metres, are common in tropical seas and feed mainly on Flying-fish which they leap after; weight up to 38 kilograms.

We have noted that some great rays jump out of the water. There are several tropical fishes which do so in various ways. The Wolf Herring, a large strap-shaped herring with cruel teeth, leaps several metres high in a semicircular arc and is no sooner back in the water than it seems to be off on another jump, followed by still other leaps in a series of huge arches.

Flying-fishes of many types are charming but shy companions of the ocean voyager. Indeed the first fish ever seen by a white man near Queensland was one of these. The French explorer Bougainville, in June 1768, noted a Flying-fish with four wings in the open sea off the Barrier Reef and said it was black with red wings. They really glide rather than fly, as they do not flap their wings. The "take-off" is accomplished through a swift zigzag movement of the tail fin after which the fish is airborne for up to 200 metres at 50 to 65 km/h.

OPEN WATER FISHES

The visitor's usual way to the Great Barrier Reef is by boat from one of the mainland ports across the expanse of water intervening. Even at the wharf his interest in fishes may be aroused by various kinds, especially Batfish, quietly browsing amongst the growths on the piles or in dense schools milling around in the shade of the wharf or the shelter of the piles, comparatively safe from attack by predacious fishes. These schooling fishes may be of several kinds at different layers (herrings, Hardyheads, anchovies, Small Yellowtail or Chanda Perches) but are most likely herrings, deeper-bodied than the more southern pilchards or sprats, but none the less related to the familiar tinned "sardine". At Murray Island, Torres Strait, these fish were exploited commercially during the 1930s. The natives used a kind of mop to drive them into nets.

As one's boat leaves port for the reef no more fish may appear unless a stop is made for handlining on the way. But the "skittering" of a Long Tom is most likely to be seen. The Long Tom is like a large Garfish but has a long upper and lower jaw with sharp teeth, rather crocodilian in appearance. Often it swims at the surface in sinuous and snaky fashion but when disturbed leaps nearly out of the water and skitters along in a series of leaps with its tail in the water and its long body at an angle of about sixty degrees to the sea; after a few such leaps it is well away and flops into safe water. Here we see the beginnings of the Flying-fish's gliding habits, though not so accomplished and lacking support from any wing-like fins. Garfishes, Long Toms and Flying-fishes are closely allied and many very young Flying-fishes possess a jutting lower jaw like the beak of a Garfish or else curious beard-like skinny appendages.

En route, a lure (either an artificial "lily", a spoon or bone jig or some red or white rag with a fair-sized hook) is trolled behind the vessel. This may be taken by one or more of quite a variety of fierce surface fishes between the mainland and the reefs. A greedy little Lizard Fish, almost as big as the bait, may snap its long-fanged jaws and hook its lizard-like head on the trap, or something more worthwhile as food may be landed. Coral Cod, usually red with a myriad blue spots, but very varied in colouring, sometimes barred, sometimes leopard-like and rarely even albino, may be hauled in. Leather-skins, Queensland Kingfish, Pike and even Tuna, as already noticed from the more open waters, may be boated during the crossing too, since these can come into the fairly sheltered shipping-route through gaps in the Great Barrier Reef, which is not, of course, a continuous wall of coral.

By handline fishing several types of sharks may be taken but more usually such attractive fish as Sweetlips (miscalled "Morwong"), Rock Cods, and those of the Bream and Emperor families are caught. The Government Bream is prized above all these. This fish is so-called because of the broad-arrow mark on each side like the old government or convict brand. In young fish there is a good deal of white between the bright sloping red lines which form the arrow and cover most of each side. Medium-sized fish are red all over except for the white belly and the arrow is darker red. Very large fish (about 1 metre long and 21 kilograms in weight) are dark red on the back and the broad-arrow is indistinguishable and these are some of the choicest fish found in Australia.

A very few fishes such as the Red Bass and Chinaman Fish, on the contrary, may be poisonous to eat, but the skipper of your boat or any local fisherman would be able to distinguish these, if appealed to, and almost all the Barrier Reef fishes may be eaten fresh with safety.

In 1770 Captain Cook's men caught on their lines a beautiful little Paradise Fish (*Pentapodus setosus*) off Bustard Bay. Solander compared its coloured sides with a rainbow and it had a long hair-like ray extending from its tail. The species is still common in our tropical waters and the most amateur fishermen have little difficulty in catching it.

The Sucking-fish is found clinging to sharks, whales, turtles, boats or anything else which may give it a free ride by means of a venetian-blind-like sucker on top of its head. It is not a Pilot Fish though often misnamed as such. Natives of Torres Strait used Sucking-fishes to catch turtles. They tied a line to the fish's tail and made it swim away from their canoe and attach itself to a turtle; this done, the line and turtle (and fish) were hauled in or a man worked his way along the line to the turtle to secure it with stronger gear. Sucking-fishes often lie under boats and swim free to take the baits intended for more desirable food-fishes.

BEACH OR SAND FISHES

Before coming to the purely coral fishes, the sand-loving species may be dealt with as they, like the kinds caught by handlining in open water, are of food and commercial value, though most of them are obtained by nets. Naturalists are always on the lookout for Lancelets, the most primitive types of fish life known. These creatures may be scooped out of sand with any improvised dredge; they are almost transparent, only a few centimetres long, and have a series of chevron-shaped muscle-bands. Only a rudimentary backbone is present, there are no jaws, the food being caught by a ring of tentacles round the mouth.

Sharks swim over the beaches at times but they are mostly of small size and too frightened of humans to attack, nevertheless a lookout should always be kept for larger man-eaters. Tiger sharks with their characteristic barred bodies and round snouts often follow one's boat to its anchorage and wait for refuse from the galley but usually large sharks only cruise in the lagoons at high water. In very shallow water the Shovel-nose Ray is very common, with its angular snout; it is quite harmless and can be caught by hand by a quick person. The Saw-fish with its toothed saw-like snout, a fearsome weapon, may come to exposed sandy beaches and can slash dangerously. The Shark Ray, which appears to be half a shark and half a ray, though it is really a ray and has curious scroll-like teeth, is harmless.

Oval-bodied Brown Rays with blue spots cloud the water with silt as they feed. When approached, they swim away, sending up puffs of silt or sand in their wake. A similarly coloured species with kite-shaped body hides in the sand with only the eyes and the banded whip-like tail showing. The Cowtail Ray swims over rocks and oyster-beds but though easily approached should not be touched as it is a Stingray like the others, with a venomous barb at the base of its tail.

The Flutemouth fish with its long, tube snout, like the old-fashioned churchwarden's pipe, sometimes swims in the shallows, changing colour from brown to striped or spotted in a chameleon fashion. Garfishes, with their long, spear-like lower jaws; Mullet, Whiting, small Long Toms, and a host of more normal-looking silvery fishes of various kinds are common inshore on sandy beaches. A haul of the seine net will reveal many species, some of which, notably Grunters, Jewfish, Toadoes, Hair-finned Trevallies and File-fishes, make an indignant grunting noise when they are caught.

Here too are small Flounders and Soles, delicious morsels, with both eyes on top of their heads as they lie on one side on the sandy floor. Most of them are too small for commercial purposes but the Queensland Halibut (with one eye on the top edge of its head) is larger and a prime food-fish. The Tongue Sole is shaped like a dog's tongue.

In the shallowest water may be seen the Darts swimming even in the little frilly waves that wash the beach. They are related to the Whitefishes and Trevallies and speaking of the latter there are many different kinds of these in Queensland, the beach-loving ones often having hair-like rays extending from the dorsal and anal fins and with silvery bodies with beautiful iridescence. Pony-fishes, with faces rather like little horses and with strongly protuberant mouths, are found with their cousins the Silverbellies, the latter having much larger scales, intensely silvery.

Feeling their way with two "barbels" which quiver in front of the chin, the Goat-fishes or Red Mullets are a feature of the sandy fauna, their "goatees" being organs of taste and feeling. The Butter-fish is silvery with dark grey spots or a combination of spots and dark bars; often wrongly called John Dories in Queensland, these fishes should be handled with care as a prick from the dorsal and other fin-spines can be painful. There are two main types of Batfish, the common silvery one which also likes wharf-piles, and the large brown one with its extensive dorsal and anal fins. These, like the Butter-fish, feed on the foulest matter in the water but their scavenging keeps the water pure and they themselves are beautiful so we need not criticize their bad taste in food.

The Silvery Moonfish, almost described by its name, is another beach fish caught by seine nets. The Spinefoot is unique amongst fishes in having more spines in the fins under the body than any other normally-shaped fish. These spines and those of the dorsal fin can give a nasty wound; and the fish, like the others just mentioned, is a scavenger, swimming in mobs or little schools. Flatheads are not numerous in shallow sandy waters of the Barrier Reef, neither are Leatherjackets, but one sometimes nets the curious Triple-spine. This is like a Silvery Leatherjacket or File-fish but has two spike-like ventral fins which form, with the tail, a sort of tripod like a photographer's, and so a dry, stuffed fish of this sort makes an interesting paperweight.

TRAWLED FISHES

The average visitor on holidays is not likely to take a trawl with him, anything more than a naturalist's dredge being very cumbersome. Yet near Lindeman Island in 1935 a few enthusiasts made a series of hauls with an otter trawl from a launch, each haul being of about an hour's duration in 5 to 7 metres of water. An average number of 113 fishes belonging to about 25 separate species rewarded this arduous hauling by hand, whether by night or day. Flounders, Angler-fishes, Pipe-fishes, Sea-horses, Goat-fishes, Dragonets and several types of food-fishes were caught but mostly of small size. Commercial trawling is not carried out to any great extent as there is so much "rough ground" due to coral growths which ruin nets. Nevertheless the Investigation Vessel *Endeavour* years ago discovered some promising food-fishes in commercial quantities. Apart from the bream-like ones there were Gurnards, Queensland Halibut, and Butterfly or

Yellow-lip Bream. The latter is a small but delicious fish with pale pink and yellow and mauve colour-pattern so sought after by the Japanese before World War II that they sent ships as far south as Australian waters to catch it. They published postcards showing its exquisite colours.

MANGROVE FISHES

The fishes of the muddy reaches of the mangrove creeks are especially numerous in Queensland. Anyone flying down the coast of the mainland must be impressed by the snake-like rivers carrying their silt to the sea with a thick shoreline or mangrove vegetation. To haul a net in these waters involves standing up to the knees at least in sucking, scratching mud, but the rewards are not only a tremendous variety of fish life but fearsome-looking stingrays and sea-snakes as well. On the islands of the Barrier Reef there are only comparatively small areas of mangrove swamp but on the mainland they are a fruitful source of the large and delicious Palmer, a fish usually called Barramundi. Other fine fishes of the mud regions are the Brown Sweetlips or Jack Johnson (a large black fish usually called after the leading negro boxer of an earlier day) and the so-called Salmon, really a member of the Threadfin family; also Soles, Bombay Duck, Catfish, Tarpon, and Lady Fish, but these are rarely encountered on the coral islands. If the tourist braves the sandflies and the oppressive sultriness, a walk through a mangrove swamp is well worthwhile since one is almost bound to see the Goggle-eyed Mangrove-fish, Kangaroo-fish, Johnny Jumper or Mud-hopper. Joseph Banks seems to have been the first white man to record it from Australia for in 1770 he mentioned it when with Cook at Thirsty Sound. The Goggle-eyed Mangrove-fish sometimes drowns if kept under water. This remarkable animal skips about over the mangrove mud or lives in burrows each with a little turret or lookout. It is credited with climbing the arched roots of mangrove trees and is often pictured in books as doing so, but recent investigations into its habits suggest that this is unusual. Aptly named Kangaroo-fish, it skips over shallow pools or quicksand-like mud with enough speed to elude pursuers. Each eye can move independently of the other and is sometimes rolled in its socket as if to moisten it.

Away from the mangroves, on rocky or stony shores, some Blennies (the Rock-hoppers or Jumping Joeys) leave the water and skip over the hard surface in a manner similar to the mud-hopper of the soft swamps.

CORAL FISHES

I have kept until last the coral-haunting fishes, most beautiful (not to say gaudy at times) of all the varied assortment of Queensland's fishes. Even the seasick passenger feels his discomfort was worth enduring when he is safely landed on a coral isle and is free to paddle in the shallow lagoon and wade in stout old boots amongst the crunching coral or better still swim with snorkel or aqualung. Here as he leaves the glistening white beach he may see several small and harmless kinds of sharks. One sort has black tips to both dorsal fins and tail but is otherwise brown; the Wobbegong is patterned like a tapestry or carpet; but most striking of all is the Catshark with its two black "eye-spots" just behind the head. This elongate shark can be picked up by the tail and merely squirms in one's hand; its mouth is too small to bite one and its eye-spots are just a bluff to frighten its enemies. When put back in the water it may "walk" away to the nearest shelter, being one of the few fishes to use its pectoral and ventral fins like arms and legs in higher vertebrates.

The wader is recommended to wear gloves because coral can give a septic scratch and it is necessary to lift rough stones and coral to see the wonders of the shore. Many fishes hide under rocks, stones or coral when the tide falls and can easily be found then. Eels, with their sharp teeth, may attack an unwary paddler and small rock cods may rush at him as if to snap his heels, but their assaults though deadly to their finny prey, are lilliputian attacks when launched upon a human being. Honeycomb Rock Cod, spotted like a giraffe, may lurk in empty clam shells or under stones. These are allied to the giant Queensland Groper, which lives in deeper water and grows to 2 metres long or more. Both the giant Groper and certain giant eels are feared by pearl divers farther north but need not worry the visitor to the average coral cay.

Under stones in shallow water may be found the rather ugly looking but harmless Frogfish with frilly side-fins. It often croaks like a frog when disturbed and the female lays very large globular eggs which are guarded by the male—a not unusual habit amongst rock-pool fishes. Here too is found the Mud Blenny or Dagger-fish, so called only on account of its shape, for it is devoid of any weapons of offence.

In very shallow water Gobies are common resting on the bottom. Each seems to have its territory and guards it against intruders. The ventral fins (under the body) are united to serve as a sucker for clinging to stones.

Striped Eel-catfish have a habit of swimming in schools, each fish conspicuously striped yellow or cream and brown; then they "go into a huddle" and form a dense ball of waving bodies and fins and seem to roll through the water rather like tumbleweed in a desert. Another fish swimming in small schools is the Razor-fish, but it swims vertically instead of horizontally and its body is thin as a wafer.

Soldier-fishes, often known as Squirrel-fishes, have very rough scales, red bodies and large saucer eyes. They are nocturnal and hide under coral shelves in the daytime. Soles are rare amongst coral except for one kind, the Peacock Sole, beautifully adorned with eye-spots like a peacock's tail. Lagoon waters are at times invaded by dense schools of Hardyheads or Atherines, in such a huddled mass that the water is dark with them and they even leap into a wader's boots to avoid the sea-birds, small sharks and other enemies which may be chasing them. They are small and sometimes mistaken for sardines but they have two little dorsal fins whereas all the fishes of the herring tribe have only one dorsal fin.

A point which strikes an observer of nature is that certain coral reef fishes have special favoured localities or even specialize to the extent of living in or with certain "hosts". Sea-horses and Pipe-fishes live, of course, amongst sea-weeds which they resemble in form and colour but if we go deeper into the subject it seems as if particular Pipe-fishes, for example, live *only* amongst certain weeds and so an evolutionist might reason that fishes and algae have evolved in partnership over many thousands, perhaps millions of years. The Painted Flutemouth lives amongst soft coral-like creatures called Gorgonias. Young Pike live in swaying sea-weeds and, when frightened, up-end themselves and sway in time with the weeds. A fish which I named the Little Fraud (*Fraudella*) is coloured like a goldfish and lives inside sponges and other growths on stones. Certain Soldier-fishes swim in and out of the water in which the needle-sharp spines of sea-urchins are waving, protected by their clever choice. Anemone-fishes are not only impervious to the stings of their hosts but nestle in the tentacles which would kill other small animals and share the anemones' meals with them and even, in some cases, with a brightly coloured prawn or two as fellow guests. The skindiver can often see large fishes quietly resting while small Blue Streak or Cleaner-fishes (*Labroides*) or prawns swim over them and pick off parasites or detritus as food. Parrot-fishes and other kinds even line up for this service, opening their mouths and gill-covers so that the little blue- and black-banded fishes can enter them, much as the crocodile bird is allowed to pick the teeth of the crocodile.

Demoiselles are very commonly seen, say through a face-mask or from a glass-bottomed boat, but certain species prefer certain types of branching coral and some (notably the "Humbugs") will not leave this coral even if it is removed from the water. The Messmate-fish par excellence however hides in the body of the trepang or sea-cucumber, occasionally coming out for a swim and returning to its unusual abode. One species enters the folds of the pearl oyster though sometimes suicides in so doing for the oyster covers the fish with mother-of-pearl, suffocates and entombs it in nacre.

The coral flourishes as a rule some distance from the shore, and in the surge-channels between the coral and down in the deep pools are to be seen those beautifully coloured fishes of many different species which rival the most brilliant birds and butterflies in coloration. The Flannelette-fish (*Rainfordia*) is rare but is gaudily striped like orange and blue flannel; certain kinds of *Pseudochromis* are so variable in colour that one fish may have a red head and a green tail while another has the very reverse and all sorts of intermediate variations are met with; little Roundheads or Imps (*Plesiops*) on the other hand are almost all black or dark brown. One kind of Dragonet is vividly patterned in straying stripes of peacock colours. But words fail to describe the multifarious and brightly-coated fishes of the corals—an inspection of George Coates's meticulous paintings made from unfaded specimens gives a much truer idea.

Smilers or Monkey-fish glide tail first into holes in the coral floor. Their huge, grinning mouths engulf the prey for which they lie in wait, but the odd thing is that no two specimens of this fish have so far been caught with identical coloration. Gold spots, blue ground-colour, brown bands—colour seems to matter little to the shy, sly Monkey-fish.

To mention all the perch-like fishes, Hussars, Emperors, bream-like species and so forth would be to compile a wearisome catalogue and no book or article could hope to cover the whole fish-fauna of the Barrier Reef. The fascination of studying them leads to further pleasures as one learns more and more. The coral fishes of popular conception are probably the Butterfly-fishes of the genus *Chaetodon,* plate-shaped, slab-sided creatures adorned with reds, blacks and yellows like the wings of butterflies. They have delicate pointed snouts for exploring the coral crannies in search of food. Some, more gorgeous than the rest, have a long spine at each cheek; these are the Angel-fishes (family Holacanthidae) and there are numerous species. The Surgeon-fishes carry a lancet-like spine on each side of the tail and can slash with these but they rarely do so; their cousins the Unicorn-fishes are more likely to do so with the blades on the butts of their tails which are as sharp as knives and dangerous to handle. Whilst the classical Unicorn-fish has a long horn on its "nose", most of the Barrier Reef species have only a small bump in this position. The most gaudily banded and quaintly outlined member of the above groups of coral fishes is the curiously named Moorish Idol. As if imitation is the sincerest form of flattery, under the sea as well as on land, the Feather-fin Bull-fish, an entirely different genus, seems to imitate to some extent the banded pattern and the pennant-like dorsal fin of the Moorish Idol.

As colourful as the macaws, lorikeets, lories and parrots in any aviary are the appropriately named Parrot-fishes. Not only have these the gorgeous greens, reds and blues of their avian counterparts but most of the Parrot-fishes have strong beaks for breaking off lumps of coral or stony weeds for food. Other Parrot-fishes, known as Wrasses, have separate, pointed teeth in the jaws, but in the Parrot-fishes proper, all the originally separate teeth have fused to form the now solid beak, which may be red, white, or blue. One Wrasse, the Jawslinger, can thrust out its jaws to an enormous extent, like a John Dory or a Pony-fish, and fold them back in front of the head again when the prey is swallowed. It swims in odd attitudes, sometimes upside down.

Even the little Gudgeons and Gobies of the coral reef are gaily coloured. Some, called Greenies, are amongst our smallest Australian vertebrates. Others are coated with a granular substance which almost hides the head and body.

Porcupine-fishes are not common but there are several kinds of Toadoes and all these are poisonous to eat. Box-fishes run riot with their colours and one kind at least has horns like a cow's and, when adult, a long, fan-shaped spotted tail. We have already noted that Leatherjackets are rare amongst coral but one small species with a very long name (*Oxymonacanthus longirostris*) is green with orange spots and another imitates a Toado in form and colour-pattern; we don't know why. Leatherjackets are to some extent replaced in the tropics by File-fishes, which are similar but have a very thick skin like the surface of a file. Many of them are most beautifully coloured; they look rather like pigs and even grunt when taken from the water.

A few coral reef fishes have discovered that it is better to lie low and not swim away. These are the Scorpion-fishes and their weird-looking relatives the Butterfly Cod and the Stonefish. The dorsal spines of all these fishes are provided with venom glands like those connected with snakes' fangs and can inflict a very painful wound; in the case of the Stonefish there may be extreme agony and even death as a direct result of the poison or through heart failure.

So much has been said on the positive side as to the variety of form, colouring and species of Barrier Reef fishes that something negative may be called for, in defence of the fishes of the southern half of Australia. Very well, then. There are no lampreys in North Queensland, no *real* Grey Nurse Sharks, and no Electric Rays, no *true* Salmon, and the Murray Cod is found only in a few mainland rivers. Though many fishes are given southern names (for instance, Snapper and Morwong), there are no Morwongs, strictly speaking, on the Barrier Reef and though the Snapper is found northward of Bowen at times, most of the "Snappers" of the reef are quite different from the well-known southern fish.

Angler-fishes, Gurnards and Leatherjackets are not as varied or plentiful as they are in the colder waters and Flatheads are fewer and different as a rule. Some of the Leatherjackets and Box-fishes of Tasmania vie in beauty of colour with coral fishes so that the warmer waters have not the monopoly of piscine beauty. In southern Australian seas, too, have developed such bizarre curiosities as the Red-Indian-fish, Goblin-fish, Leafy Sea Dragon and others which have never been found on the Great Barrier Reef. It is, however, dangerous to be dogmatic, especially in science. Nature is in a continuous state of flux and ebb and flow. So many exceptions to our arbitrary rules occur that it would be rash to say that such and such a fish does not or cannot live on the Great Barrier Reef—an exception is likely to turn up and disprove the statement.

A great deal has still to be found out and put on record concerning the growth, migrations, breeding, eggs, food and habits of the Great Barrier Reef fishes and any careful observations, especially if supported by preserved specimens, are of scientific and practical value. Considering all this, we are impressed anew with the truth of Aesop's saying:

What is discovered only serves to shew
That nothing's known to what is yet to know.

Description of the Fishes

Family Heterodontidae: Port Jackson Sharks

Port Jackson Shark *Heterodontus portusjacksoni* **PLATE 74:1**

Enlarged lateral teeth without distinct keels. Colour light brownish or greyish with blackish harness-like stripes on the body and a broad blackish bar below the eye. Found in southern waters of Queensland. Grows to 1.4 metres. Harmless. Feeds upon molluscs, etc.

Crested Port Jackson Shark *Heterodontus galeatus* **PLATE 74:2**

Enlarged lateral teeth with distinct keels. Colour light brownish; back with some dark transverse bars but not joining to form a "harness" as in the previous species. A southern shark which enters the waters of southern Queensland. It grows to 1.2 metres. Harmless.

Family Carcharhinidae: True Sharks

Tiger Shark *Galeocerdo cuvier* **PLATE 75:1-3**

Young with dark blue tiger-like stripes which fade out in the adults. A fierce and dangerous shark which grows to a length of 6 metres or more and inhabits the outside waters, bays and rivers of Queensland. Known by its blunt head and its oblique serrated strong teeth.

Weasel Shark *Hemigaleus microstoma* **PLATE 75:4**

Teeth serrated and oblique in the upper jaw only, lower ones much smaller, narrower and not serrated. Colour silvery-grey, white below. A small shark which inhabits the waters of North Queensland. It grows to a length of 70 centimetres or so and is considered harmless.

School Shark *Galeorhinus australis* **PLATE 76:1-3**

Teeth acute, depressed, coarsely serrated. Colour slaty-purplish, bluish-grey or brownish above, much lighter below; fins without dark marks. It enters Moreton Bay, southern Queensland from southern waters. Grows to over 1.75 metres. Not considered dangerous to man.

White-tip Shark *Triaenodon obesus* PLATE 76:4

Teeth similar in both jaws, backwardly directed, each with three entire acute cusps. Colour greyish or brownish; dorsal and caudal fins with milky-white tips. A shark found in the waters of northern Queensland. It is said to grow to a length of 1.75 metres and is most probably dangerous.

Blackspot Shark *Carcharhinus sealei* PLATE 76:5

Teeth in lower jaw slender, curved, entire, and sloping outwards, those in upper jaw may have the cusps minutely serrate or entire. Colour satiny-grey; a dark brown blotch on half of second dorsal. A small shark known only from North Queensland. Grows to 80 centimetres.

Dusky Shark *Carcharhinus obscurus* PLATE 77:1

Teeth serrated in both jaws, upper ones triangular, oblique laterally, lower ones erect and smaller. Colour sandy or dark grey. Inhabits the outside waters of Queensland and also enters the bays and rivers. Grows to a length of 3.5 metres at least. Dangerous.

Bronze Whaler *Carcharhinus brachyura* PLATE 77:2

Teeth in both jaws with broad bases, those in lower jaw with the cusps very narrow; very much broader in upper jaw but considerably narrower than in true Whaler, *Eulamia macrura.* Inhabits the open sea off the coasts of southern Queensland and New South Wales. Grows to about 3 metres and 320 kilograms.

Graceful Shark *Carcharhinus amblyrhynchoides* PLATE 77:3

Teeth compressed, erect, and serrated in both jaws, those in the lower jaw more minutely so. Colour coppery, with a bluish tinge above and on fins; white below. Common in the northern waters of Queensland. Grows to at least 60 centimetres, probably much more.

Black-tip Shark *Carcharhinus melanopterus* PLATE 77:4

Teeth finely serrated with age, edges nearly or quite smooth in the young. Readily recognized by the black blotches on the tips of the fins. A common shark and a pest to line fishermen in Queensland. Grows to a maximum length of about 2 metres and could be dangerous to man.

Estuary Shark *Carcharhinus leucas* **PLATE 77:5**

Teeth in both jaws triangular and serrated, those in the upper jaw broad, lower ones erect and narrower. Colour greyish to bluish above, lighter below. A dangerous shark found along the coast and in the estuaries and rivers of Queensland. It grows to at least 2 metres.

Little Blue Shark *Scoliodon laticaudus* **PLATE 77:6**

Teeth in jaws similar, oblique, their outer bases finely serrated, the cusps smooth. Colour bluish-grey or grey-brown. One of the most common of the small sharks on the Queensland coast. It has been trawled down to 73 fathoms in the south. Harmless. Grows to 1.25 metres.

Smooth-fanged Shark *Carcharhinus brevipinna* **PLATE 78:1**

Teeth entire, those in lower jaw smaller, more narrow and erect than those in the upper jaw. Colour greyish-blue or greyish-brown above, white below; an indistinct band along middle of body. A small shark found in the northern waters of Queensland. Grows to 80 centimetres.

Lemon Shark *Negaprion acutidens* **PLATE 78:2**

Teeth entire, bases broad, smooth on both base and cusp; the cusps erect, narrow, and lanceolate. Colour greyish or brownish above, whitish below. This Queensland shark grows to over 1.75 metres in length. The young ones move about over the many reefs in schools.

Taylor's Shark *Rhizoprionodon taylori* **PLATE 78:3**

Teeth in upper jaw with straight tips, those in the lower jaw with tips bent slightly inwards. Colour bluish-grey or fawnish, paler below, the two colours fairly sharply defined. A shark commonly taken in the mackerel nets in southern Queensland. Grows to 83 centimetres.

Slender Dog Shark *Loxodon macrorhinus* **PLATE 78:4**

Teeth quite smooth on both base and cusp. Colour dark bluish-grey above, grey below; pectoral and ventral fins with ashy-grey outer edges. A small and fairly common Queensland shark from the inshore waters down to 25 fathoms or so. Grows to about 90 centimetres.

Gummy Shark *Mustelus antarcticus* **PLATE 78:5,6**

Teeth in jaws smooth, blunt, flattened, in pavement-like formation. Colour ashy-grey above, white below, the two colours fairly sharply demarcated posteriorly. The flesh of this shark is excellent eating and the livers yield valuable oil. Length about 1 metre. Southern Queensland.

Family Sphyrnidae: Hammer-head Sharks

Hammer-head Shark *Sphyrna lewini* **PLATE 79:1**

Eyes situated at the tips of greatly expanded lobes on each side of the head, the lobes functioning as a "bow-rudder". The young frequent the bays and rivers, the adults wandering the open ocean and hunting in packs. Grows to 6 metres. A dangerous shark to man.

Family Orectolobidae: Wobbegongs and Cat Sharks

Tawny Shark *Nebrius concolor* **PLATES 79:2, 81:5**

Teeth small, in three series and multicuspid. Colour sandy-brown or tawny. A common and harmless shark which has been reported from various points along the Queensland coast. Fishermen say that when captured it coughs and spits water. It grows to least 3 metres.

Tasselled Wobbegong *Orectolobus ogilbyi* **PLATE 79:3**

Teeth dissimilar, very sharp. Almost a continuous row of branched skinny lobes on sides of head. Colour light sandy, covered with a network of dark brown rings. A shark found in the northern waters of Queensland. It grows to 3 metres and could be dangerous to man.

Spotted Wobbegong *Orectolobus maculatus* **PLATE 79:4**

Teeth fang-like, entire. Sides of head with only a few widely spaced lobes. Colour brown; head, back, and fins spotted and marked with pale brown and creamy rings, arcs, or blotches. Found in the southern waters of Queensland. Grows to 3 metres. Probably dangerous.

Banded Wobbegong *Orectolobus ornatus* **PLATE 79:5,6**

Teeth fang-like and dissimilar. Colour brown, marbled darker and lighter to form obscure dark brown cross-bars with creamy eye-like spots. A common shark on the Queensland coast, usually among weed-covered rocks and in the coral growths. Grows to a length of at least 2 metres.

Northern Wobbegong *Orectolobus wardi* PLATE 80:1

Nasal fringes and dermal lobes simple. Ground colour yellowish, back greyish; three large, dark, white-edged spots before first dorsal fin; dark bars on sides; tail with three dark bands; fins with dark spots. Found in the northern waters of Queensland. Grows to 45 centimetres.

Blind Shark *Brachaelurus waddi* PLATE 80:2

Teeth small, tricuspid, entire. Colour dark brown to blackish, sometimes with white spots and flecks; yellowish below; juveniles have dark transverse bands. It is not blind but takes its name from its habit of closing the eyelids when taken from water. Length 90 centimetres.

Blue-grey Cat Shark *Brachaelurus colcloughi* PLATE 80:3

Teeth small, tricuspid. Anal fin only slightly behind the second dorsal. Colour ashy-grey above, passing into white below. This rare shark, known only from a single example, was captured at Mud Island, Moreton Bay, southern Queensland, in 1908. Its length is 45 centimetres.

Epaulette Shark *Hemiscyllium ocellatum* PLATE 80:4

Teeth small, tricuspid. Colour rich brown above, the back, sides, and fins with large blackish-brown spots; brownish-white below; the denticles of the skin highly polished; a large eye-like spot just behind last gill-slit. Common on coral reefs. Length 90 centimetres. Harmless.

Speckled Cat Shark *Hemiscyllium trispeculare* PLATE 80:5

The colour is similar to that of *H. ocellatum,* but this shark is covered by small brownish dots which are absent on the under-surface and between the tail cross-bands. A small and harmless shark inhabiting northern reefs of Queensland. Grows to 60 centimetres.

Brown-banded Cat Shark *Chiloscyllium punctatum* PLATE 81:1-3

No eye-like spot above the pectoral fin. Colour of adults is uniform reddish-brown or greyish-fawn; some examples have small, dark brown spots; young are brownish, with 10 or more dark cross-bands. Whole of Queensland coast. Grows to length of at least 1 metre. Harmless.

Zebra Shark *Stegostoma varium* PLATE 81:4,6,7

Colour: Juveniles are dark brown or black above, with creamy-white bands and spots on head, back, and tail. Adults are tawny or creamy-yellowish with dark brown spots over head, body, and tail. Found along whole Queensland coastline. Length 3.25 metres. A harmless shark.

Family Alopiidae: Thresher Sharks

Thresher Shark *Alopias vulpinus* PLATE 82:1

Teeth triangular, edges smooth. Tail longer than length of body. Colour grey or bluish-grey above, white below, the junction of the two colours sometimes with dark blotches. Grows to 5.5 metres. The long tail is used to frighten schools of fishes upon which it feeds.

Family Carchariidae: Sand Sharks

Grey Nurse *Odontaspis taurus* PLATE 82:2

Snout pointed; mouth wide, curve almost pointed in front. Teeth long, slender, awl-like. Colour varies from light grey to pale brownish above, dirty white below. Grows to at least 4.5 metres. Commonly taken more in the waters of southern Queensland. A dangerous shark.

Family Isuridae: Porbeagles or Mackerel Sharks

Blue Pointer or Mako *Isurus oxyrinchus* PLATE 82:3

Body torpedo-shaped, snout pointed; a keel on each side of the tail. Teeth long and awl-like. Colour dark blue above, white below, the two colours sharply demarcated. A savage, dangerous shark, not commonly found in Queensland waters. A pelagic species which grows to 4.5 metres.

White Pointer *Carcharodon carcharias* PLATE 83:1,2

Body spindle-shaped, snout pointed, mouth large; teeth large, erect, regularly triangular, serrated. Colour pale grey or slaty-brown above, shading to white below. A powerful and most dangerous shark; grows to 12 metres and a weight of over 2 tonnes. It inhabits most warm open seas.

Family Pristidae: Saw Fishes

Narrow-snouted Saw-fish *Pristis zijsron* **PLATE 84:1,2**

Snout produced into a long, heavy, flattened, saw-like weapon armed with "teeth" on either side. Colour greenish-grey, becoming yellowish on sides; whitish below. Grows to over 7 metres. Common. Harmless, unless one is accidentally struck by the "saw". Feeds upon small sea creatures.

Family Rhinobatidae: Shark Rays, Shovel-nose Rays and Fiddler Rays

Shark Ray *Rhina ancylostomus* **PLATE 84:3,4**

Snout broad, semicircular; mouth deeply undulated. Teeth in pavement-like bands. Head and shoulders armed with stout bony bucklers. Colour dull brown, lighter below. Sometimes with dots and spots. Not common in Queensland. Grows to 2.75 metres and 124 kilograms in weight.

White-spotted Ray *Rhynchobatus djiddensis* **PLATE 84:5**

Body depressed. Teeth pavement-like, surfaces undulated. Back with fairly large tubercles. Colour greyish or brownish, with white spots or ocelli; a black spot present on scapular region. Fairly common in Queensland. Length 3 metres; weight 226 kilograms. Harmless. Flesh edible.

Common Shovel-nose Ray *Rhinobatos batillum* **PLATE 85:1,4**

Nostrils very oblique; mouth almost straight. Colour light brownish or olive, white below; pectorals and ventrals yellowish. A common Queensland ray which freely enters both brackish and fresh waters. It grows to 2 metres and is a good food-fish in its young stages. Harmless.

Banks' Shovel-nose Ray *Aptychotrema rostrata* **PLATE 85:2,3**

Nostrils almost transverse; mouth slightly curved. Colour yellowish or olive, fins lighter, under-surface white. The common shovel-nose of Moreton Bay, where it frequents sandy beaches. Also taken by trawl down to 30 fathoms. Grows to a length of 1.25 metres. Flesh of good flavour.

PLATE 85:5

Bougainville's Shovel-nose Ray *Aptychotrema bougainvillii*

Nostrils more transverse than oblique; mouth strongly curved and bow-shaped. Colour brownish, with a few marks on the fins. This ray is occasionally taken in the waters of southern Queensland. It grows to about 83 centimetres and is quite harmless. A good edible species.

Fiddler Ray *Trygonorrhina fasciata* PLATE 85:6

Colour light brown, with a bluish-grey pattern bordered with dark brown or black and in shape somewhat similar to the scroll-like holes in a violin, hence the name of Fiddler Ray. A southern ray occasionally taken in Moreton Bay, southern Queensland. Grows to 1.25 metres. Harmless.

Family Torpedinidae: Electric Rays

Numb-fish *Hypnos monopterygium* PLATE 85:7

Tail short; eyes small. Colour dark brown above, white below. Has large internal electric organs between head and pectorals. Is capable of giving severe electric shocks when touched. Common in the southern waters of Queensland. Harmless unless touched. Grows to 60 centimetres.

Family Rajidae: Skates

Common Skate *Raja australis* PLATE 86:1

Tail slender, depressed, but not whip-like and without a spine; skin more or less spiny. Colour olive-brownish or dark greenish above; light spots or markings may be present. A deep-water skate trawled in southern Queensland. Grows to 1.25 metres. It is a valued food-fish.

Family Dasyatidae: Stingrays

Blue-spotted Stingaree *Amphotistius kuhlii* PLATE 1:1,2

Tail long, whip-like, with at least one serrated spine. Young are pale grey or brown, spotted with black, or rusty-red and white; adults have large bluish eye-like spots, the black spots of the young having been lost. Common on reefs of North Queensland. Grows to width of 25 centimetres.

Estuary Stingaree *Dasyatis fluviorum* **PLATE 86:2**

Tail long, whip-like, with one or two serrated spines; a broad patch of tubercles on shoulders. Colour olive-brown. Extremely common in the bays and estuaries of southern Queensland. Grows to a width of 1.25 metres. Many rays are dangerous to handle on account of their serrated spines.

Cow-tail or Fantail Ray *Dasyatis sephen* **PLATE 86:4**

Tail fairly long, with serrated spine and a broad cutaneous fold below, which extends about half-way to its tip. Lead-colour or blackish above, white below. A common Queensland ray seen on shallow flats when the tide is rising. It grows to a width of at least 1.75 metres.

Mangrove Ray *Himantura granulata* **PLATE 86:3**

Tail long, with one or more serrated spines; entire disc covered with granules. Colour uniform brown, sometimes with a few scattered white spots. A common Queensland ray found lying in shallow waters, usually among mangroves. Grows to a width of at least 33 centimetres. Flesh edible.

Coach-whip Ray *Himantura uarnak* **PLATES 2:1, 86:5**

Tail long, with at least one serrated spine; granules in mostly two to three widely separated rows on head. Colour brown, with darker spots; tail with black and white rings. Very variable in colour. A common ray in Queensland. Grows to width of 1.5 metres. Its ringed tail identifies it.

Blue-spotted Lagoon Ray *Taeniura lymma* **PLATE 2:2**

Tail compressed, longer than body, with one or two spines. Colour tan-brown above, with large bright blue spots; a broad blue stripe along each side of tail. A handsome ray very common in the northern waters of Queensland. Reaches a width of about 38 centimetres. Flesh quite edible.

Common Stingaree *Urolophus testaceus* **PLATE 87:1**

Tail short, stout, muscular, with spine and terminating in a well developed caudal fin. Colour reddish or brownish, extreme edges white; white below. A common Queensland ray seen on sandy flats in shallow water but also enters deep water. Grows to width of about 33 centimetres.

Rat-tailed Ray *Gymnura australis* PLATE 87:2

Tail short, thin, with serrated spine; disc wider than long, becoming more so with age. Colour brown or greenish-grey, variegated lighter and darker or speckled with small black spots. Fairly common along the Queensland coast. Grows to a width of at least 1 metre. Edible.

Thorny Ray *Urogymnus asperrimus* PLATE 87:3

Body covered with tubercular scales varying in size, shape, and numbers and with numerous erect thorns. Colour greenish or brownish above, yellowish to whitish on tubercles. Rare, but has been recorded in northern Queensland waters. Grows to a width of at least 60 centimetres.

Family Myliobatidae: Eagle Rays

Spotted Eagle Ray *Aetobatus narinari* PLATE 87:4

Head prominent, snout produced; tail long, with one or more serrated spines. Colour greyish, greenish-olive, or lead colour above, with bluish-white spots. Grows to a width of 3.25 metres and weight of up to 200 kilograms. A very destructive ray where oysters are farmed.

Eagle Ray *Myliobatus australis* PLATE 87:5

Head prominent, snout narrower forward; tail long, with serrated spine. Colour greenish, olive, or sandy, with indistinct bluish blotches, sometimes bright yellow. A pest on oyster-banks in southern Queensland. It crushes the shells with powerful teeth. Width to 1.25 metres.

Cow-nose Ray *Rhinoptera neglecta* PLATE 87:6

Head prominent, snout in two separate lobes; tail long, slender, with serrated spine. Colour uniform brown above, white below. Not commonly met with in Queensland. Has been taken near Bundaberg and also in Moreton Bay, southern Queensland. Grows to width of about 1 metre.

Family Mobulidae: Devil Rays or Sea Devils

Australian Devil Ray *Manta alfredi* **PLATE 88:1-4**

Head with cephalic fins (not curled outwards); mouth anterior. Colour dark bluish-black or slaty-grey above, white below, with dull blue-black blotches. Frequents rocky headlands in winter. Feeds upon plankton and small sea creatures. Grows to a width of at least 4.5 metres. It is harmless.

Pygmy Devil Ray *Mobula diabolus* **PLATE 88:5**

Head with cephalic fins (curled outwards); mouth inferior, well beneath head. Colour greyish or grey-brown above, lighter below; pectorals blackish terminally. A common Queensland ray seen moving in small schools in shallow water. Grows to a width of about 2.5 metres.

Family Elopidae: Tarpons

Giant Herring *Elops hawaiiensis* **PLATE 89:1**

Colour brilliant silvery, bluish on back; fins yellowish, tinged blackish, except pectorals, which are black, their inner surfaces white. A large Queensland fish which grows to over 1 metre but is of no value as a food-fish, its flesh being tasteless and full of bones.

Family Megalopidae

Ox-eye Tarpon *Megalops cyprinoides* **PLATE 3:1**

Colour bluish-green above, sides silvery, white below; fins tinged yellowish. The eastern representative of the famous American Tarpon. Whole of Queensland coastal waters, the young entering rivers into the fresh water. Grows to 1.5 metres. The flesh is of poor quality and bony.

Family Albulidae

Bone-fish *Albula vulpes* **PLATE 3:2**

Colour silvery; margins of dorsal and caudal dusky, other fins clear; some examples have dusky lines along the back. An active and beautiful game-fish found along the Queensland coast. Freely enters estuaries. Grows to about 1 metre but is of no value as a food-fish.

Family Chanidae

Salmon Herring *Chanos chanos* PLATE 89:2

A brilliantly silver fish, swift in action. Found along whole of Queensland coastline. It freely enters freshwater streams and lakes and is both a good game-fish and food-fish. It grows to 1.75 metres. It is farmed by the natives of the Philippines in huge tide-ponds.

Family Clupeidae: Herring-like Fishes

Wolf Herring *Chirocentrus dorab* PLATE 3:3

Colour deep bluish-green above, silvery on sides and below; a golden stripe separates the two colours; dorsal and caudal yellowish, edged blackish; pectorals dusky; anal clear. Noted for its leaping powers. It has been known to clear a travelling launch. Of no edible value.

Whiskered Anchovy *Thryssa setirostris* PLATE 3:4

Colour dirty greenish above, silvery below; black small-veins on the shoulders; pectorals and caudal yellowish, the latter edged black. The remarkable extension of the upper jaw distinguishes this fish. Found in the northern waters of Queensland. Grows to 20 centimetres. Edible.

Hamilton's Anchovy *Thryssa hamiltoni* PLATE 4:1

Colour greenish-grey above, silvery on sides and below; black small-veins at shoulders; pectorals yellowish; ventrals and anal whitish; caudal yellowish, edged black. Found along the shores and in the estuaries of Queensland in great shoals. Grows to 20 centimetres. Edible.

Estuary Anchovy *Thrissina baelama* PLATE 89:3

Colour buff on back and upper sides down to a line from eye to tail; lower sides and belly silvery, and pinkish; usually a shoulder spot; dorsal and caudal buff, tipped blackish; other fins white. A common Queensland anchovy often caught in river-nets. Grows to 15 centimetres.

Gulf Anchovy *Stolephorus carpentariae* PLATE 89:4

Colour orange, with a broad silvery stripe. Probably occurs along the whole coast of Queensland, as it has been found in the Gulf of Carpentaria in the far north and also in the far south of the State in Moreton Bay. Lives in large shoals. Grows to 5 centimetres. Edible.

De Vis' Anchovy *Stolephorus devisi* **PLATE 4:2**
Colour greyish-white, with a silvery lateral band along the sides; fins clear, but the tail is tinged orange. Found in shoals along the whole coastline of Queensland. Grows to a length of 13 centimetres. Edible. Such anchovies are destined to become of great economic importance.

Australian Anchovy *Engraulis australis* **PLATE 89:5**
Colour greenish above, silvery below; a broad stripe along side. Mainly frequents the southern waters of Queensland in vast shoals during winter months. It is a very good food-fish so far not utilized. Closely related to the European anchovy. It grows to about 13 centimetres.

Hairback Herring *Nematalosa come* **PLATE 4:3**
Colour greenish-grey or bluish-grey, silvery below; six to seven dark, longitudinal lines along centres of scales on back; black shoulder spot present. A common mud-eating fish which lives in coastal waters, estuaries and freshwater streams of Queensland. Grows to 25 centimetres.

Blue Sprat *Spratelloides delicatulus* **PLATE 89:6**
Colour is intense dark blue, with the silver sides cleanly divided; eye silver. Visits the shores of Queensland in vast shoals during the winter months. Its red flesh is of very fine flavour and it could be of great economic importance some day. It grows to 10 centimetres.

Hasselt's Sprat *Dussumieria hasseltii* **PLATE 5:1**
Colour beautiful dark green, shot with blue above; silvery below; under the blue is a bronze line, with a silver one beneath; pectorals, ventrals and anal white. A large sprat found in northern waters of Queensland. It grows to 20 centimetres and is a good food-fish.

Freshwater Herring *Potamalosa richmondia* **PLATE 90:1**
Colour dark green above, sides silvery, white below. A herring which inhabits freshwater streams of northern New South Wales, coming down into the sea for the spawning season. It has been recorded from southern

Queensland only occasionally. A good food-fish. Grows to 25 centimetres.

Sandy Sprat *Hyperlophus vittatus* PLATE 90:2
Colour light green, with a broad, well-defined silvery stripe on side; belly white; dorsal and caudal fins yellow. Visits the waters of southern Queensland in vast shoals during the autumn and winter months. A good food-fish, but little used. It grows to 13 centimetres.

Glassy Sprat *Hyperlophus translucidus* PLATE 90:3
Body is translucent in life, with a broad silvery band on sides; back with scattered black dots. Occasionally enters Queensland waters from the south, extending only as far north as Caloundra, southern Queensland. It grows to 5 centimetres and could be a valuable food-fish.

Spotted Herring *Harengula koningsbergeri* PLATE 5:2
Colour greyish-green above, silvery below; a salmon-coloured stripe along upper sides is overlaid with a row of dusky bars or spots and sometimes a few below. A common herring along the coastline of North Queensland. Could be of economic importance. Grows to 18 centimetres.

Southern Herring *Harengula castelnaui* PLATE 90:4
Colour blue above, silvery on sides and below; back with one to three more or less distinct golden, dark-edged bands; dorsal and caudal tinged with black on edge. A common herring on the coast and in the estuaries and inlets of Queensland. Moves in great shoals. Grows to 23 centimetres.

Dakin's Pilchard *Sardinella sirm* PLATE 90:5
Colour blue-grey above, yellowish or silvery below; about 11 more or less distinct dark spots along the upper sides. A common Torres Strait pilchard found in vast shoals in far northern waters of Queensland. Grows to 18 centimetres. A valuable unused food-fish.

Australian Pilchard *Sardinops neopilchardus* PLATE 5:3
Colour dark blue-greenish, changing abruptly to silver on sides; each scale on back with a small black spot; another row along junction of blue and silver. Common. Destined to be of commercial importance. Occurs in vast shoals in southern Queensland. It grows to 28 centimetres.

Ditchelee Herring *Pellona ditchela* **PLATE 90:6**

Colour silvery, with burnished lateral band; a dusky shoulder spot present; dorsal and caudal fins with dark margins, other fins clear. This herring is not common in Queensland but it has been trawled in the northern part of the State. It grows to a length of 20 centimetres.

Family Gonorhynchidae

Rat-fish *Gonorhynchus greyi* **PLATE 91:2**

Colour greyish or brownish above, silvery below; a pink stripe runs from head to tail; fin-tips black; tail with black patch on each lobe and the tips pink. Frequents sandy bottoms, in shallow water, but also has been found at considerable depths. Length 45 centimetres. Edible.

Family Aulopidae

Sergeant Baker *Aulopus purpurissatus* **PLATE 91:1**

Second and third dorsal rays produced into a long filament in the males. Colour mottled purplish or pinkish-red; scale edges crimson and crimson blotches on back and sides; fins yellow, with red spots or bands. Southern waters of Queensland. Grows to 60 centimetres. Edible.

Family Synodontidae: Grinners or Lizard-fishes

Painted Grinner *Trachinocephalus myops* **PLATE 91:3**

Colour golden-brown above, yellowish below, back with more or less distinct dark cross-bands; sides with faint bluish lines, edged dark brown; a black shoulder blotch. Teeth visible when mouth is closed. Found on Queensland coast. Flesh full of small bones. Length 33 centimetres.

Hoult's Grinner *Synodus houlti* **PLATE 91:4**

Colour greyish, sides and belly white; back with ill-defined faint cross-bars, more defined at tail base; fins without markings. A fish which is not met with by the casual observer. Has been trawled in Queensland down to 30 fathoms. Grows to at least 20 centimetres.

Spear-toothed Grinner *Synodus sageneus* PLATE 91:5

Colour lilac, with narrow longitudinal pencil lines above; a purplish lateral band divides the lilac from the yellowish-white below. A rare fish which has been found on the east coast of Queensland and also in the Gulf of Carpentaria. It grows to a length of 26 centimetres.

Yellow-banded Grinner *Saurida tumbil* PLATE 91:6

Colour of head and body pale reddish-brown, mottled darker; silvery below; fins dusky creamy-yellowish, dotted with black; dorsal and caudal more dusky. Common in Queensland waters. It grows to 45 centimetres. Grinners inhabit most warm seas, the species being very numerous.

Family Harpodontidae: Bummalows, "Bombay Ducks"

Glassy Bombay Duck *Harpodon translucens* PLATE 92:1

Young are transparent in life, with minute black specklings. Adults tinged greenish above, opalescent-pink below; fins dusky-tipped; everywhere minute specklings. A flabby fish with bristle-like barbed teeth. Taken in the estuaries of Queensland. Length 40 centimetres.

Family Plotosidae: Catfishes

Striped Catfish *Plotosus lineatus* PLATE 92:2

Colour brown above, lighter below; fins dark; second dorsal, anal, and caudal with dark edge; young are olive-greenish, with two or three yellowish stripes along sides. Common in the sea, estuaries, and rivers of Queensland. Must be handled with care. Grows to 30 centimetres.

White-lipped Eel Catfish *Paraplotosus albilabris* PLATE 92:3

Colour uniform blackish; under-surface of head and abdomen greyish; lips white. First dorsal fin almost as high as the body. A common fish in the coastal waters of Queensland. It should be handled with care because of its dorsal and pectoral spines. Grows to about 35 centimetres.

Estuary Catfish *Cnidoglanis macrocephalus* PLATE 92:4

Colour uniform blackish-brown above, whitish below; fawn under head, faintly freckled, darker; pectorals and ventrals dusky above, lighter below. A common fish inhabiting muddy bottoms along the Australian coastline but not commonly seen in Queensland. Grows to 90 centimetres.

Nettle-fish *Euristhmus nudiceps* **PLATE 92:5**

Colour light brown or fawnish, the fins with black edges. A species from the Arafura Sea. It is fairly commonly met with in Queensland as far south as the Whitsunday Passage. It grows to at least 30 centimetres and, like others of the family, should be handled very carefully.

Long-tailed Catfish *Euristhmus lepturus* **PLATE 92:7**

Colour brownish or dark greyish-brown above, lighter below and obscurely mottled with lighter areas on tail; soft dorsal and anal with dark borders. Some examples have creamy-white spots on body and tail. Common in estuarine waters of southern Queensland. Length 35 centimetres.

Family Ariidae: Sea Catfishes

Giant Salmon Catfish *Arius thalassina* **PLATE 92:8**

Colour reddish-brown above, whitish below; fins almost clear; dorsal, adipose, anal, and caudal fins dusky at tips. A giant catfish found in the northern waters of Queensland where it lives in the estuaries. Grows to a length of 1 metre and is a good food-fish.

Family Gobiesocidae: Shore Eels

Red-banded Pygmy Eel or Shore Eel *Alabes dorsalis* **PLATE 93:1**

Colour green, with darker spots when alive, but when placed in preservative it changes to red, hence the former name of *rufus*. A common southern pygmy eel recorded in southern Queensland from a few examples only. It is usually found just below the tide-mark. Grows to 10 centimetres.

Family Anguillidae: Freshwater Eels

Long-finned Freshwater Eel *Anguilla reinhardtii* **PLATE 93:2**

Colour olive-greenish above, sides golden-yellowish, more or less spotted and streaked blackish; white below; fins darker. Common in most coastal streams of Queensland, from salt to fresh water. It is a valuable food-fish which grows to 1.25 metres and a weight of 6.5 kilograms.

Family Muraenesocidae: Pike Eels

Pike Eel *Muraenesox bagio* PLATE 93:3

Colour greyish above, sides silvery; fins with a broad black margin. A very vicious eel found in the sea and also in brackish and freshwater streams of Queensland. The powerful teeth can inflict terrible bites. It will attack when taken into a boat. Grows to at least 1.5 metres.

Family Congridae: Conger Eels

Australian Conger Eel *Conger verreauxi* PLATE 93:4

Colour brownish-yellow, with purplish tinges; white or yellow below; edges of the vertical fins darker. A large and powerful eel which is very rarely found in Queensland. It grows to a length of at least 1.5 metres and a weight of 17 kilograms. Very vicious and will readily attack.

Family Muraenidae: Moray or Reef Eels

Clouded Reef Eel *Echidna nebulosa* PLATE 93:5

Colour yellowish or whitish, with a row of large black blotches which are broken up into star-like spots. A pretty eel commonly seen among the corals on the Great Barrier Reef. It grows to 75 centimetres and like all reef eels must be handled carefully to avoid severe bites.

Painted Reef Eel *Gymnothorax pictus* PLATE 95:1

Young are yellow, with round blackish spots about the size of the eye and in three longitudinal rows; adults are brownish-grey, with very small black spots. Very variable. A beautiful eel, commonly seen gliding among the coral growths on the reefs of Queensland. Length to 1.25 metres.

Tessellated Reef Eel *Gymnothorax favagineus* PLATE 95:2

Colour blue-black, covered with white lines which form a tessellated pattern; the young have the white lines much wider, giving the effect of large black spots. A common, vicious eel found along the whole Queensland coastline among coral growths and rocks. Length 1.25 metres.

Long-tailed Eel *Evenchelys macrurus* **PLATE 95:3**

Colour is uniform brownish-grey. Identified by its greatly attenuated form, its tail being twice as long as the body. Frequents rivers, ascending into the fresh waters. The world's record is of one taken in the Maroochy River, Queensland. Its length was 3.93 metres.

Family Ophichthidae: Worm and Snake Eels

Iredale's Worm Eel *Muraenichthys iredalei* **PLATE 95:4**

Colour brownish, darker on the back, where there are numerous blackish punctations; the viscera shows through just behind gill-slits. This small worm eel is found mainly on the reefs of North Queensland. It grows only to a length of about 16 centimetres. Not commonly met with.

Half-banded Snake Eel *Leiuranus semicinctus* **PLATE 95:5**

Colour yellowish-brown, the body adorned with black saddle-like bands; dorsal and anal fins not confluent, the caudal wanting. A handsome eel found on the coral reefs and inshore in Queensland. It grows to at least 50 centimetres. It can "back" swiftly into the corals.

Family Belonidae: Long Tom or Needle-fishes

Black-finned Long Tom *Tylosurus acus* **PLATE 96:1**

Colour intense green above, silvery-white below, divided by a broad silvery blue-green band. Readily recognized by the large compressed bony appendage below the lower jaw tip. Rare in Queensland, its inclusion based on one taken in Moreton Bay. It grows to about 1 metre.

Black-spot Long Tom *Strongylura strongylura* **PLATE 5:4**

Colour brownish or yellow-green above, sides silvery, white below; a silvery blue-bordered band along posterior half of body; dorsal, anal, and caudal yellow, the latter with a black spot; pectorals and ventrals clear. Common. Edible, but extremely bony. Grows to 45 centimetres.

Stout Long Tom *Lhotskia gavialoides* PLATE 96:2

Colour green above, sides silvery, white below; dorsal and caudal fins green, with dusky borders; outer half of pectorals black; ventrals yellowish. This is the common Long Tom of southern Queensland. Apparently not met with in the north. Grows to 1.25 metres. Edible, but very bony.

Barred Long Tom *Ablennes hians* PLATE 96:3

Colour dark brownish, bluish, or greenish, sides and below silvery and white; a narrow dark green vertebral stripe from head to dorsal fin, which is a broad sea-green band to the tail; dark cross-bands on body posteriorly; inside mouth scarlet. Rarely seen. Length 1.25 metres.

Family Hemirhamphidae: Garfishes or Half-beaks

Long-jawed Garfish *Rhynchorhamphus georgii* PLATE 96:4,5

Greatly developed fleshy flaps below and on each side of lower jaw. This unique fish, about 13 centimetres in length, was collected in 1924 at Townsville, North Queensland. Apparently it has not been met with since. Probably silvery in life; the beak and folds are black.

Short-nosed Garfish *Hyporhamphus quoyi* PLATE 96:6

Colour yellowish-green, with tints of blue and pink; scales on back flecked golden-brown and bright green; a silvery band on the side. A valuable food-fish found along the whole coastline of Queensland. It grows to a length of 35 centimetres and is very common.

Black-barred Garfish *Hemirhamphus far* PLATE 6:1

Colour greenish-blue above, sides silvery, with four to nine large black spots along sides; tail with upper lobe yellow, lower lobe blue. This beautiful large species is common along the coast of Queensland. It grows to 50 centimetres and is a valuable food-fish. Flesh excellent.

Three-by-two Garfish *Hemirhamphus robustus* PLATE 6:2

Body robust, sides flattened. Colour dark green above, silvery on the sides, with a silvery band; a black spot usually present on each side below dorsal fin. Found along whole Queensland coast. Grows to 40 centimetres. A good food-fish. Identified by its flattened sides.

Spoon-fin Garfish *Zenarchopterus dispar* **PLATES 6:3, 97:1-3**
Colour olive-brown above, sides paler, with a silver band with blue-black line above it; whitish below; fins dusky. Frequents the sea, as well as the rivers and mangrove creeks of Queensland, where it floats in schools on the surface in still waters. Grows to 15 centimetres.

Long-finned Garfish *Euleptorhamphus viridis* **PLATE 97:4**
Colour light bluish-silvery above, scales darker-edged; sides and belly silvery; upper lobe of tail with dusky band along edge. An open-sea fish very seldom taken in Queensland. The two which have been recorded here both jumped aboard travelling launches. Grows to 45 centimetres.

Snub-nosed Garfish *Arrhamphus sclerolepis* **PLATE 6:4**
Colour sea-green above, with three fine black lines from head to dorsal fin; sides pinkish; a bluish stripe along side, a silvery one below. A common Queensland food-fish. Although found in the sea, is also a lover of freshwater streams. Commonly found far inland. Grows to 38 centimetres.

Family Exocoetidae: Flying-fishes

Short-finned Flying-fish *Parexocoetus brachypterus* **PLATE 97:5**
Colour dark blue above, silvery below; a dark patch on dorsal fin; pectorals densely covered with small black spots; upper lobe of tail blue-grey, lower lobe dark blue. Known by its short pectorals and black patch on the dorsal. Rarely taken in Queensland. Grows to 20 centimetres.

Common Flying-fish *Exocoetus volitans* **PLATE 97:6**
Colour bluish above, silvery on sides; pectorals bluish-black or brownish, broadly white posteriorly and with upper and lower borders white; other fins pale. This fish is readily recognized by its long pectorals. Outside waters of Queensland. Grows to 30 centimetres. Edible.

Flying-fish *Hirundichthys oxycephalus* PLATE 97:7

Colour dark bluish above, silvery below; pectorals dusky, with a lighter transverse band; middle rays of ventrals and caudal fin dusky. As with most species of flying-fishes, it is seldom met with owing to its open-sea habits. It grows to a length of 23 centimetres.

Family Aulostomidae: Trumpet-fishes

Spiny-back Trumpet-fish *Aulostomus chinensis* PLATE 98:1

The colour of this fish is very variable. Usually uniform brown, with or without cross-bars or lines. Some examples are yellow. A widely distributed fish which is not commonly found on the coast of Queensland. It grows to a length of at least 76 centimetres.

Family Fistularidae: Flute-mouths or Cornet-fishes

Flute-mouth *Fistularia commersonii* PLATE 6:5

Colour greenish-brown, whitish below; two interrupted blue stripes from head to tail. Some have bright orange fins. They feed upon small fishes and, before attack, they float in the tide as motionless and straight as a stick. Common in Queensland. Grow to a length of 1.5 metres.

Family Macroramphosidae: Bellows-fishes

Long-spined Bellows-fish *Macroramphosus elevatus* PLATE 98:2

The colour of this curious fish is orange-red. It has been trawled from the deep water off the coast of southern Queensland down to Victoria and Tasmania and also New Zealand. The body, which is covered with bony plates, is more or less rigid. It grows to about 20 centimetres.

Family Centriscidae: Razor-fishes

Razor-fish *Centriscus cristatus* PLATE 98:3

Colour silvery, with a deep red band from snout to eye; a golden spot on the temple, followed by a red or orange band which runs to the terminal spine. Fairly common on the Queensland coast. Swims vertically in shoals usually with the head down. Body rigid. Grows to 30 centimetres.

Family Syngnathidae: Pipe-fishes and Sea-horses

Blue-striped Pipe-fish *Doryhamphus melanopleura* **PLATE 98:4**

Colour bright orange-red, with ultramarine-blue longitudinal stripe from the end of the snout to the tail, which is bright blue, orange basally and with white edges. This handsome pipe-fish inhabits the coral reefs of North Queensland. It grows to about 6 centimetres.

Short-tailed Pipe-fish *Microphis manadensis* **PLATE 98:5**

Colour dark olive-brown above, paler and profusely spotted with brown below; sides of the snout with about nine conspicuous black spots; tail black. A wide-ranging species which, although recorded from Moreton Bay, is apparently rare in Queensland. It grows to 25 centimetres.

Black-chinned Pipe-fish *Syngnathus altirostris* **PLATE 98:6,7**

Colour uniform brown, with a dark stripe from tip of snout through eye to lower half of gill-cover, where it breaks into blotches. A pipe-fish from southern Queensland and northern New South Wales entering and also living in fresh water. It grows to a length of about 15 centimetres.

Messmate Pipe-fish *Corythoichthys intestinalis* **PLATE 99:1**

Colour greyish, with numerous dark cross-bars or with dark blotches along the sides; a black central streak on the throat. This pipe-fish is said to live in the intestinal cavity of the bêche-de-mer, a type of sea-slug. Inhabits Queensland waters. Also wide-ranging elsewhere.

Thorn-tailed Pipe-fish *Micrognathus brevirostris* **PLATE 99:2**

Colour variable; usually dark brown with lighter cross-bars on each ring across the back; sometimes with a brown- and pearl-coloured spot on the gill-cover. Has been recorded from Cape York Peninsula in far North Queensland. It grows to a length of about 7.5 centimetres.

Gray's Pipe-fish *Halicampus grayi* **PLATE 99:3**

Colour brownish with whitish blotches; sometimes the gill-cover has white stripes. A rather distinctive wide-ranging little species which has been recorded from the waters of Torres Strait, North Queensland. A pipe-fish not commonly met with. It grows to 15 centimetres.

Banded Pipe-fish *Yozia tigris* **PLATE 99:4**

Colour dark green above, brownish on sides, white below; a few narrow, reddish-brown stripes on sides of head; body with broad dark reddish bands; a white spot on lower edge of each body scute. Found on the Queensland coast from north to south. It grows to 30 centimetres.

Pig Pipe-fish *Choeroichthys brachysoma* **PLATE 99:5**

Colour brown; light and dark pattern on snout, chin, and throat; dark bar through eye; row of reddish spots along upper sides anteriorly; three pairs of blotches across back; tail dusky-brown, hind third yellowish. Found in the waters of North Queensland. It grows to about 5 centimetres.

Spotted Pipe-fish *Yozia bicoarctata* **PLATE 99:6**

Colour in preservative brownish, with a yellow spot on each body ring below the lateral line. A little-known pipe-fish from the waters of Torres Strait, North Queensland. It grows to at least 28 centimetres in length and is apparently rare in our waters.

Wide-bodied Pipe-fish *Stigmatophora nigra* **PLATE 100:1**

Pale green, speckled with minute black dots; lower surface of snout with two dark longitudinal bars extending back behind the eye. Body of male scarcely broader than deep. Body of female strongly depressed, in the adults twice as broad as deep. Southern Queensland. Length 10 centimetres.

Horned Pipe-fish *Syngnathoides biaculeatus* **PLATE 100:2**

Colour pale green or brown, with or without dark brown spots. A poor swimmer, inhabiting shallow waters all along the Queensland coast. It attaches itself by its prehensile tail to marine growths. Said to be used in China as a medicine. It grows to at least 25 centimetres.

Red and Gold Sea-horse *Solegnathus dunckeri* **PLATE 100:3**

Colour reddish, darkest on tail; a blackish stripe along the scutes forming the dorsal ridges. A large and handsome fish found along the Queensland coast. It grows to 45 centimetres and is used by natives in the islands as a file. Extremely rough and tough. Files wood easily.

Slender Tangle-fish *Haliichthys taeniophora* **PLATE 100:4**

Colour brown, with irregular dark brown bands across the back; abdomen whitish; cutaneous appendages black. Ranges from North Queensland and southern New Guinea, across North Australia to Western Australia. The one illustrated lacks the cutaneous appendages. Grows to 30 centimetres.

Common Sea-horse *Hippocampus whitei* **PLATE 100:5**

Colour very variable; usually brownish or olive-brownish, with the snout, head, and body marbled with brown; sometimes yellowish, with small red spots. A very common sea-horse, found in the southern waters of Queensland. It grows to a length of about 15 centimetres.

Low-crowned Sea-horse *Hippocampus planifrons* **PLATE 100:6**

Colour olive or orange-yellow, everywhere with irregular bright orange spots, with numerous whitish dots between; some have dark brownish zebra-like stripes. Found along the Queensland coast and commonly taken in the prawn-trawls in Moreton Bay. Grows to 10 centimetres.

Family Trachipteridae: Ribbon-fishes

Southern Ribbon-fish *Trachipterus jacksonensis* **PLATE 101:1**

Colour bright silvery, with a pink dorsal fin. A rare southern fish which has been taken in southern Queensland. One specimen, found on the beach at Southport in November 1921, measured 1.98 metres. Mostly known from cast-up ones on beaches. Appears to live in deep waters.

Family Regalecidae: Oar-fishes

Oar-fish *Regalecus pacificus* **PLATE 101:2**

Colour brilliant silvery, tinged lavender-grey; dorsal with red rays, the membranes rose-red; anterior half of body with black, vertical streaks, followed by rows of oval, blackish spots. Recorded from southern Queensland through two specimens cast up on our beaches. Grows to 25 centimetres.

Family Caproidae

Roseate Boarfish *Antigonia rubicunda* PLATE 101:3

Colour rose-pink, with a darker band from nape through eye to throat; a second one from dorsal spines to pectoral and a third one around tail base. A rare fish known only from two examples trawled in the Capricorns in 70 fathoms. They measured about 6 centimetres.

Family Monocentridae: Knight-fishes

Knight-fish *Cleidopus gloriamaris* PLATE 101:4

Colour yellow, edges of the body scales black, forming a lattice pattern. Some scales with orange bases; lips jet-black, the lower one with scarlet patch and brilliant white line above, this being an organ of luminescence that glows at night. Common. Grows to 23 centimetres.

Family Berycidae: Nannygai

Nannygai *Centroberyx affinis* PLATE 101:5

Colour bright red, with reflections of purple and gold above, the sides paler, with centres of scales darker, forming longitudinal bands; fins with rays red, membranes pink; tail red, with yellow edge. Enters Moreton Bay occasionally. A good food-fish. Grows to 45 centimetres.

Family Trachichthyidae: Roughies

Roughy *Trachichthys australis* PLATE 102:1

Colour rich reddish-brown; a dark vertical band on the gill-cover; all fins, with the exception of pectorals, with dark brown areas; an almost colourless yellowish-tinged band at base of dorsal and anal fins. A southern fish which straggles into Moreton Bay. Grows to 15 centimetres.

Slender Roughy *Optivus elongatus* **PLATE 102:2**

A black band along each side of the caudal lobes and another in front of the soft dorsal and anal fins. This rare deep-water fish was recorded in Queensland from a single example which was trawled in Moreton Bay, southern Queensland, in 1889. It grows to about 13 centimetres.

Family Holocentridae: Soldier-fishes

Scarlet-fin Soldier-fish *Adioryx spinifer* **PLATE 7:1**

Colour pinkish-red, each scale with a silvery centre; a few violet dots along junctions of some lateral scale-rows; a spot of crimson behind the eye, another at pectoral axil; first dorsal scarlet. Inhabits the Great Barrier Reef of North Queensland. Grows to about 38 centimetres.

Crowned Soldier-fish *Adioryx diadema* **PLATE 7:2**

Colour reddish, with white longitudinal stripes; spinous dorsal scarlet, with white membranes anteriorly. Apparently uncommon in waters of Queensland, where it lives among reefs. Grows to 23 centimetres. Scales of soldier-fishes have needle-like points. The heads are spiny.

Red and White Soldier-fish *Adioryx cornutum* **PLATE 102:3**

Scarlet-red bands on a white background; spinous dorsal scarlet-red and black; black spot at tail-base and sometimes one on soft dorsal and anal. A strikingly handsome fish, common on the numerous coral reefs of North Queensland. It grows to a length of about 16 centimetres.

Red Soldier-fish *Adioryx ruber* **PLATE 8:1**

Colour red, with silvery-white longitudinal stripes along the rows of scales; fins rosy or yellowish; some black in the anal fin and also on upper and lower margins of the tail. A common soldier-fish in Queensland and seems to prefer the deeper waters. Grows to about 28 centimetres.

Blunt-nosed Soldier-fish *Myripristis hexagonus* **PLATE 102:4**

Colour rose-pink, darkest on head, each scale edged darker pink; a large white patch behind the pupil of the eye, which is surrounded scarlet; edge of the gill-cover dark red on upper half. A common fish among the coral reefs in both deep and shallow water. It grows to about 19 centimetres.

Family Zeidae: John Dories

John Dory *Zeus faber* PLATE 103:1

Colour dull greenish or greenish-brown, silvery below; a large blackish patch, bordered silvery, situated about the middle of the side. An uncommon fish in Queensland, where it straggles into Moreton Bay from the south. Mostly taken by trawlers. Excellent food-fish. Grows to 60 centimetres.

Family Kuhliidae: Flag-tails

Rock Flag-tail *Kuhlia rupestris* PLATE 8:2

Colour bright silvery, each scale strongly outlined in black, but the outline lost on the pure silver below; tail blackish, pale yellowish terminally, ventrals white; anal pale yellow. Occasionally found in the sea but mainly inhabits the freshwater streams of Queensland. A sporting and food-fish; grows to 45 centimetres.

Family Apogonidae: Cardinal-fishes

Black-finned Cardinal-fish *Apogon nigripinnis* PLATE 9:1

Colour dark brown; a large, yellow-edged spot above the pectoral fin; lower half of head and throat, also pectorals and tail, pale yellow. A handsome Queensland species, mostly taken by the prawn-trawlers when capturing prawns in the otter-trawls. Length to about 10 centimetres.

Cardinal-fish *Apogon cooki* PLATE 9:2

Colour olivaceous or pearly-grey, with three broad dark olive-green longitudinal stripes, usually ending in a dark spot at the tail-base; fins pinkish or brick-red, a dark band at base of soft dorsal and anal fins. Swims in shoals among the branching corals. Common. Length 13 centimetres.

Bar-cheeked Cardinal-fish *Apogon bandanensis* PLATE 103:3

Colour brown, with brassy reflections; an obscure brown, oblique streak below eye; three dusky-brown, obscure, saddle-like patches on back. A variable species, common on the coral reefs of North Queensland. It is readily recognized by the streak below the eye. Grows to 10 centimetres.

Siphon-fish *Siphamia cuneiceps* **PLATE 103:4,5**

Colour brassy, everywhere covered with minute black dots; fins clear, except ventrals and anal, which are blackish on their basal half. Commonly taken by otter-trawl in southern Queensland. Grows to about 5 centimetres. Both sides of the body with a tube-like prolongation.

Rose-bellied Siphon-fish *Siphamia roseigaster* **PLATE 103:2**

Colour silvery-white above, passing into pale rose-colour below, the tube-like prolongation on each side of the body being the colour of quicksilver. Common in the harbours and estuaries of southern Queensland, where it is captured in prawn scoops at night. Length about 7.5 centimetres.

Five-striped Percelle *Cheilodipterus quinquelineatus* **PLATE 104:1**

Colour of head and body pale pink, with five dark, horizontal stripes from head to tail, the centre one ending in a black, yellow-encircled spot at the tail-base; tail pink, with dark upper and lower edges. Common in the coral reefs of North Queensland. Grows to about 13 centimetres.

Family Plesiopidae: Long-fins

Blue-spotted Long-fin *Plesiops coeruleolineatus* **PLATE 104:2**

Colour blackish or very dark brown, each scale with a blue spot; similar spots on the head and shoulders; a large black blue-edged spot on the gill-cover; pectoral margin red. More commonly met with in the southern half of Queensland. Grows to about 20 centimetres. Very pretty.

Blue-tip Long-fin *Paraplesiops poweri* **PLATE 104:3**

Colour purplish-black or greenish-brown, with faint greyish transverse bands; a blue band below the eye and blue cheek-spots; fins purple, edged with blue; pectorals greenish-yellow. Found along the Queensland coast but far more common southwards. Grows to 20 centimetres.

Family Priacanthidae: Bulls-eyes or Big-eyes

Red Bulls-eye *Priacanthus macracanthus* **PLATE 10:1**

Colour red, silvery below; vertical fins red or pink, with dark spots on the membranes; pupil of eye black, surrounded by scarlet; inside mouth orange-red. Queensland from north to south, young frequenting shallow bays in great numbers in winter months. Grows to 30 centimetres.

Family Toxotidae: Archer-fishes

Rifle-fish or Archer-fish *Toxotes chatareus* PLATE 10:2

Colour silvery, with golden flushes, greenish-brown dorsally; six to seven blackish spots of unequal size between eye and end of dorsal. Young have larger spots. North Queensland, in brackish and fresh waters, where trees overhang. Shoots insects with a jet of water. Grows to 30 centimetres.

Family Centropomidae

Sand Bass *Psammoperca waigiensis* PLATE 104:4

Colour dull brown, centres of scales darker, fins brownish, except pectorals and ventrals which are yellowish-white; eye reddish, like a live coal. Inhabits the Queensland coast from north to south. Frequents reefs and rocky areas. A good food-fish which grows to about 35 centimetres.

Giant Perch or Palmer *Lates calcarifer* PLATE 104:5

Colour of adults greenish-grey, sides silvery, white below; head with purplish tinges; spinous dorsal with spines silvery, membranes dark; eye brilliant pinkish-red. Inhabits the sea, tidal waters and rivers of Queensland north from the Mary River. Valued food-fish. Grows to 1.75 metres.

Yellow Perchlet *Priopodichthys marianus* PLATE 105:1

Colour whitish, upper scales margined with olive-green spots; a thin dark line on median line of tail. A marine and estuarine species found along the whole coastline of Queensland. It grows to about 11 centimetres. It is an admirable aquarium fish and is very active and hardy.

Family Glaucosomidae

Pearl Perch *Glaucosoma scapulare* PLATE 105:2

Colour silvery greenish-brown above, silvery below, with numerous small indistinct golden-brown spots; a prominent supraclavicle is deep iridescent blue-black. One of the best Australian food-fishes. It frequents rocky grounds and reefs of southern Queensland. Length 60 centimetres.

Family Serranidae: Sea Basses and Rock Cods

Yellow Emperor *Diploprion bifasciatum* PLATE 11:1

Colour clear lemon-yellow, with two broad brownish or blackish cross-bands, one from spinous dorsal, through eye to throat, the other, much wider, across the body from dorsal to anal. A very handsome fish not commonly found in Queensland waters. It grows to at least 30 centimetres.

Lunar-tailed Rock Cod *Variola louti* PLATE 11:2

Colour of head, body and fins brownish-red, darkest on back and becoming reddish-orange below, everywhere with spots of crimson-lake, lilac, or purple; fins edged yellow. A good game and food-fish commonly caught over the coral reefs of Queensland. Grows to at least 76 centimetres.

Blue-spotted Rock Cod *Cephalopholis cyanostigma* PLATE 12:1

Colour brownish-red, with about six indistinct mesh-like bands, and covered with black-edged blue spots; fins edged with blue; tips of spinous dorsal and inside lips scarlet. Commonly inhabits the coral reefs of North Queensland. A good food-fish. It grows to 35 centimetres.

Freckled Rock Cod *Cephalopholis sexmaculatus* PLATE 12:2

Colour scarlet, covered with blue spots; six black saddle-like patches along the back; pectorals orange; ventrals and anal narrowly edged blue. A handsome fish known only from the reefs of North Queensland. Grows to at least 76 centimetres and a weight of 6.5 kilograms.

Coral Trout *Cephalopholis miniatus* PLATE 12:3

Colour scarlet above, orange-scarlet below, covered with small, black-edged blue spots; pectorals orange-yellow; fins and tail with dark blue borders. A very pretty fish, common on the reefs north of and including the Capricorns. Flesh of excellent flavour. Length 45 centimetres.

Yellow-spotted Rock Cod *Epinephelus areolatus* PLATE 13:1

Colour reddish-brown or greenish-brown, with darker or yellow hexagonal spots or markings formed by bluish-white lines on head, body and fins, the latter with dark margins edged with white. A common fish on the coral reefs of North Queensland. It grows to about 40 centimetres.

Maori Rock Cod *Epinephelus undulatostriatus* PLATE 105:3

Colour very light creamy-brown or stone-grey, with numerous narrow umber or golden-brown stripes, obliquely upwards above lateral line and more or less horizontal below; head spotted with gold; fins tipped with gold and with gold spots and edged with yellow. Common. Grows to 60 centimetres.

Black-tipped Rock Cod *Epinephelus fasciatus* PLATE 13:2

Colour scarlet, with darker vertical bands; fins scarlet, passing into orange-yellow; membranes between dorsal spines deep sepia in the upper part; a black circle around the eye and black above edge of mouth. A very common and good food-fish in Queensland. Grows to 38 centimetres.

Estuary Rock Cod *Epinephelus tauvina* PLATE 13:3

Colour very variable. Young are dark olive-green, with rust-red spots and obscure bands or blotches on body; adults almost uniform dark brown, the spots indistinct. A common fish in Queensland. It grows to at least 2 metres and 226 kilograms and could be dangerous to man.

Groper *Epinephelus lanceolatus* PLATE 14:1

Young are golden-yellow, with broad dark brown or blackish bands or reticulations; adults are uniform brown or blackish, with fins and tail dirty yellowish. The largest fish in Queensland waters. It grows to 3 metres and a weight of at least 270 kilograms. It is dangerous to man.

Honeycomb Rock Cod *Epinephelus merra* PLATE 14:2

Entire head and body covered with brownish, more or less hexagonal spots, separated by a network of pale yellowish lines. Variable in colour and pattern. A good food-fish, common on the coral reefs of Queensland. Hides under dead coral at low tide. Grows to 45 centimetres.

Coral Cod or Leopard-fish *Plectropomus maculatus* **PLATE 14:3**

Colour varies greatly, being salmon-pink, reddish, or light or dark brown, covered with numerous small round, black-edged, blue spots; spots may be less numerous, larger, and elongate. Common over the coral reefs. Takes a trolling line. Length to at least 1.25 metres. Good food-fish.

White-lined Rock Cod *Anyperodon leucogrammicus* **PLATE 15:1**

Colour brownish or olive-greenish above, creamy-white below; three to five bluish or white longitudinal lines on body; head and body spotted all over with orange-rufous or reddish spots and to a lesser degree on fins. Occasionally taken over the northern reefs. Length to 50 centimetres.

Hump-backed Rock Cod *Cromileptes altivelis* **PLATE 15:2**

Colour fawn, reddish-brown or terra-cotta, covered all over with round blackish spots, larger and less numerous in the young. Common on the northern reefs of Queensland but rarely taken in the south. It is a good food-fish and grows to a length of about 66 centimetres.

Pretty-fins *Centrogenys vaigiensis* **PLATE 105:4**

Colour light fawn over head and body, with large patches of dark brown and varying spots of brown and grey; eye encircled with alternate brown and cream, like wheel-spokes. Inhabits the Queensland coast. Bears a striking resemblance to the scorpaenoid fishes. Length 15 centimetres.

Jameson's Rock Cod *Ellerkeldia jamesoni* **PLATE 105:5**

Colour dark olive-brown above, grey below; grey lines along back and irregular, transverse, slightly oblique reddish-brown bands below the lateral line; fins tinged with pinkish-red. A small fish found among rocks and reefs in southern Queensland. It grows to 10 centimetres.

Short-finned Cichlops *Dampieria lineata* **PLATE 106:1**

Colour (in spirits) is a rich brownish-yellow; a blue line almost encircles the eye and similar lines run obliquely forward and downward on the cheek; each body scale has a basal dark spot with a blue centre. An uncommon fish in Queensland waters. Grows to a length of 25 centimetres.

PLATE 106:2

New Holland Cichlops *Pseudochromis novaehollandiae*

Colour variable; some are dark brown, tinged with red, or head and foreparts of body deep pink, the rest sage-green; a blue line around lower and hind margin of eye, also on cheeks; dorsal green, margined yellow and blue. Inhabits Queensland reefs. Length to about 18 centimetres.

Family Branchiostegidae

Ward's Horse-head *Branchiostegus wardi* PLATE 106:3

General colour in formalin is light brownish above, white below; some yellow on head, pectoral bases and axilla; faint orange or yellow blotches along sides; tail yellowish. A rare southern species which has been trawled in 50 fathoms in South Queensland. Length to 44 centimetres.

Family Cepolidae: Band-fishes

Band-fish *Cepola australis* PLATE 106:4

Colour of body and fins bright red or pink, passing into pale pink below; a row of orange-yellow spots along sides; cheek with a pink patch; eye surrounded by scarlet. Occasionally captured in the trawls of the prawning boats operating in Queensland. It grows to 45 centimetres.

Family Teraponidae: Grunters or Perch

Crescent Perch *Terapon jarbua* PLATE 16:1

Colour brownish-grey above, silvery below, with dark longitudinal curved bands along back and sides; spinous dorsal with a large black blotch; other dusky spots on soft dorsal; tail with oblique bars. A common Queensland fish which grows to 28 centimetres. An edible species.

Trumpeter Perch *Pelates quadrilineatus* PLATE 16:2

Colour drab-brown or dusky greenish above, whitish below; sides with deep brown longitudinal straight bands; usually a large dark shoulder spot. Young have additional broad vertical bands. A common fish on the Queensland coastline over sand-flats. Grows to 20 centimetres.

Six-lined Perch *Pelates sexlineatus* PLATE 106:5

Colour greyish or bluish above; silvery below, with dark bands longitudinally; dorsal fins with the upper portions black; tail with blackish margin. In the young there are usually three dark blotches on the back. A common fish in North Queensland waters. Grows to 26 centimetres.

Family Sillaginidae: Whitings

Trumpeter Whiting *Sillago maculata* PLATE 106:6

Colour sandy-brown above, sides silvery, white below; a silver band on each side; back and sides with dusky blotches; dorsals spotted; pectorals yellow, with a black basal spot; anal yellowish; tail greyish. A valuable food-fish commonly taken in Queensland. Grows to 30 centimetres.

Sand Whiting *Sillago ciliata* PLATE 106:7

Colour pale brown above or olive-brown; cheeks with golden reflections; dorsal fins spotted brown; ventrals and anal golden; the pectorals with a dark blotch at base. One of the most important food-fishes of Queensland. Grows to 45 centimetres in length and over 1 kilogram.

Family Coryphaenidae: "Dolphins"

"Dolphin" *Coryphaena hippurus* PLATE 16:3

Colour brilliant blue or green above, shading into silvery-white or bright yellow, with purple and gold reflections and with spots of blue-green; eye emerald-green around the black pupil. An open-sea and fast swimming predatory fish. Grows to 1.75 metres and at least 28 kilograms.

Family Rachycentridae

Sergeant-fish *Rachycentron canadus* PLATE 107:1,2

Colour in adults is deep brown, sharply defined by a narrow silvery longitudinal band; another brown band below the silvery one; young are more strongly banded. Tail in young is truncate, in adult crescent. A game-fish which reaches 1.75 metres and a weight of at least 67 kilograms.

Family Pomatomidae

Tailor *Pomatomus saltatrix* PLATE 107:3

Colour sea-green, sides silvery, white below; fins green, tinged yellowish. A powerful, fast-swimming, ferocious fish, the young hunting in shoals, the adults in small mobs. Feeds largely upon mullet. A good food-fish but flesh soft. Common. Grows to 1.25 metres and 12 kilograms in weight.

Family Monodactylidae

Diamond-fish *Monodactylus argenteus* PLATE 17:1

Colour silvery-green, with purplish tints; dorsal and anal lobes blackish; tail yellow, with narrow black edge; young with a narrow brownish band through the eye and another through pectoral-base to belly. Frequents wharf-piles. Common. Grows to only 23 centimetres but has flesh of excellent flavour.

Family Pempheridae

Steelback *Leptobrama mülleri* PLATE 17:2

Colour silvery; back steel-blue, the two colours sharply defined; a black blotch on dorsal lobe; tail yellowish-green; anal tinged yellow. A common Queensland fish which grows to 43 centimetres and a weight of at least 1 kilogram. Its flesh is of excellent flavour. Frequents coastline.

Family Kyphosidae: Blackfishes and Drummers

Blackfish *Girella tricuspidata* PLATE 107:4

Colour grey or brownish-grey, white below; back and sides with dark, narrow, transverse bands; cheeks, lips, and throat tinged yellow. Found in the southern rivers, estuaries and coastal waters of Queensland. A good food-fish but should be skinned before cooking. Grows to 71 centimetres.

Topsail Drummer *Kyphosus cinerascens* PLATE 107:5

Colour silvery-grey, with golden bands on sides; fins dusky. Found on the Queensland coast from north to south. Its flesh is of good edible quality.

The species is readily recognized by its high dorsal fin. It grows to a length of at least 45 centimetres.

Family Lutjanidae

Chinaman-fish *Symphorus nematophorus* **PLATE 18:1,2**

Adults crimson-pink, with obscure transverse bands; head with fine yellow lines. Young are roseate, with yellowish tinge and wavy bands of blue-green longitudinally. Common in North Queensland. Grow to 90 centimetres and 15 kilograms. At times some have very poisonous flesh.

Mangrove Jack *Lutjanus argentimaculatus* **PLATE 19:1**

Colour reddish-brown, cherry-red or pinkish-red; sometimes scales on sides have pearly centres or a dark basal spot, forming lines. Found in the sea, rivers, and mangrove-lined creeks of Queensland. A good sporting and food-fish which grows to 90 centimetres and 12 kilograms weight.

Blue-banded Sea Perch *Lutjanus kasmira* **PLATE 19:2**

Colour of head reddish-brown or bluish above, yellow below and on body, fins, and tail; longitudinal blue lines from head to tail. It inhabits the reef waters of Queensland and apparently does not exceed a length of about 38 centimetres. It is quite a good little food-fish.

Red Emperor *Lutjanus sebae* **PLATE 20**

Young pale pink, with three vivid red transverse bands forming a "broad arrow". Adults salmon-pink, with a white spot on each scale base. An excellent food-fish commonly met with over the Queensland reefs and in the deeper outside waters. Grows to at least 1.25 metres and 20 kilograms.

Yellow-banded Hussar *Lutjanus amabilis* **PLATE 21:1**

Colour rose-pink, darkest on head above; a brilliant lemon-yellow band runs from head to tail. A handsome fish, common in the Capricorns but preferring the deeper outside waters farther northwards. It is a good food-fish of an average length of 30 centimetres but attains 45 centimetres.

Scarlet Sea Perch *Lutjanus malabaricus* **PLATE 21:2**

Colour deep red, with narrow yellow lines following the scale rows; a blaze of scarlet from snout through eye to dorsal origin; a dusky saddle over caudal base. Inhabits the reefs of North Queensland and according to Coates it attains a weight of 13 kilograms. Good food-fish.

Black-spot Sea Perch *Lutjanus fulviflamma* PLATE 22:1

Colour very variable; may be olive-green, purplish-red, or pink, but mostly with the longitudinal orange or yellow bands and the black spot below the dorsals. Inhabits the Queensland coast, the young entering mangrove creeks. Common. It grows to 43 kilograms. Quite a good food-fish.

Kelp Sea Perch *Lutjanus bohar* PLATE 22:2

Colour victoria-lake above, passing into rose-doree on sides and pink below. A beautiful fish fairly common on the outside reefs of North Queensland. Some have been known to cause severe illness when eaten. Grows to at least 79 centimetres and 9 kilograms.

Red-bellied Fusilier *Caesio erythrogaster* PLATE 22:3

Colour blue above, passing into rose-red below; tail bright yellow, the three colours blending below the soft dorsal. A common and very beautiful fish found among coral reefs of North Queensland. Although small it has flesh of good flavour. It grows to about 35 centimetres.

Family Nemipteridae

Pearly Spine-cheek *Scolopsis margaritifer* PLATE 23:1

Colour dark olive-brown above, each scale with a pearly-green centre, forming a chequered pattern which becomes a series of yellow spots on the white belly. A common little fish on the coral reefs of North Queensland. Grows to 25 centimetres. Each cheek armed with a strong spine.

Blue-banded Whiptail *Pentapodus setosus* PLATE 23:2

Colour olivaceous above, whitish or yellowish below; snout with blue and yellow cross-bands; two yellow and blue bands commence at the eye and run above and below the pectoral fin to the tail-base. A common little fish on the Queensland coast. It grows to 25 centimetres.

Theodore's Butterfly Bream *Nemipterus theodorei* PLATE 107:6

Colour lilac-pink above, mauve on sides, pearly iridescent pink below, with pale greenish-yellow longitudinal stripes; tail pale rose-pink, upper lobe dusky-tipped. A pretty fish known only from a few examples trawled off Caloundra, southern Queensland. Length 25 centimetres.

Family Haemulidae

Spotted Javelin-fish *Pomadasys hasta* PLATE 24:1

Colour olivaceous above, silvery on sides, white below, each scale with a dark olive-brown spot, forming more or less irregular rows; sometimes the spots are replaced by transverse bands above. A good food-fish, commonly found on the Queensland coast. Grows to 60 centimetres.

Brown Sweet-lips *Plectorhynchus nigrus* PLATE 108:1

Colour is usually dark brown, blackish or greyish, a little paler on the sides; fins blackish. Variable in colour. A common fish in Queensland waters, especially in the north. It grows to a length of about 60 centimetres and it has flesh of fair edible quality.

Painted Sweet-lips *Plectorhynchus pictus* PLATE 24:2

Young whitish or yellowish, with chestnut-brown or blackish bands longitudinally from head to tail. In half-grown examples the bands break into rows of spots, the adults being silvery-greyish with spots on body and fins. Common Queensland fish growing to 66 centimetres. Edible.

Family Lethrinidae: Emperors

Yellow-tailed Emperor *Lethrinus mahsena* PLATE 25:1

Colour greenish, white below, scales outlined in brown; upper lip salmon, scarlet in gape; fins edged red; pectoral rays orange-rufous, membranes hyaline, first ray blue; tail yellow-edged. Variable in colour. Common on the coral reefs of North Queensland. Grows to 45 centimetres.

Spangled Emperor *Lethrinus nebulosus* PLATE 25:2

Colour pale olive-green or fawnish, white below, each scale with a pearly-blue central spot and a much smaller basal spot; cheeks with blue bands and spots; inside mouth red; fins and tail pinkish with pale spots. A reef-dwelling food-fish of Queensland. Length to 76 centimetres.

Sweet-lip Emperor *Lethrinus chrysostomus* PLATE 26:1

Colour olive-greenish, lighter below, with darker transverse bands; head blood-red between eyes and on to snout; red in mouth; dorsal fins with scarlet membranes. The most common *Lethrinus* of the Queensland coast. It is a splendid food-fish, growing to 90 centimetres and 9 kilograms.

Red-finned Emperor *Lethrinus fletus* PLATE 26:2

Colour fawnish-olive above, lighter below; pale pink horizontal lines along scale-rows; blue lines and spots on head; inside mouth scarlet. Sometimes body is blotched. Common along the Queensland coast. It is a good food-fish, growing to 45 centimetres. Average about 33 centimetres.

Family Lobotidae

Triple-tail *Lobotes surinamensis* PLATE 108:2

Colour umber or blackish, darker on head; paler below, sometimes yellowish; pectorals and margin of tail pale yellowish. Attains to 1 metre and a weight of 13 kilograms. Coastal waters of Queensland. When netted exhibits remarkable jumping powers. Flesh of good quality.

Family Sparidae: Snapper and Sea Bream

Collared Sea Bream *Gymnocranius audleyi* PLATE 26:3

Colour silvery; pectorals yellowish; ventrals whitish; fins narrowly edged red; whitish "collar" above and behind eye. Although found along the whole Queensland coastline this fish is far more common north of the Capricorns. A good food-fish. Grows to 58 centimetres and 4 kilograms.

Cape Moreton Sea Bream *Dentex spariformis* PLATE 108:3

Colour pink above, washed with gold; silvery below; head dull red above; dorsal and caudal fins saffron-yellow, the base and lateral borders of the latter pink; upper half of pectorals yellow. Trawled in deep water off Cape

Moreton, southern Queensland. Grows to about 25 centimetres.

Snapper *Chrysophrys auratus* PLATES 21:1, 180:4

Colour pink, silvery-white below, with numerous bright blue spots. Very young fish have dark transverse bands. Young prefer sheltered bays and inlets. Adults frequent outside deeper waters, travelling in schools. A valuable food-fish, growing to 1.3 metres and 20 kilograms.

Long-spined Snapper *Argyrops spinifer* PLATE 27:2

Colour reddish or rose-pink above, silvery below; pink lines along the scale rows; young with broad transverse bands across body; fins pink. It frequents the Queensland coastline over sunken shoals and reefs. A good food-fish but rather uncommon. It grows to 60 centimetres.

Tarwhine *Rhabdosargus sarba* PLATE 28:1

Colour silvery-greyish, each scale with a gold centre, forming longitudinal lines on body; the pectorals and ventrals are yellow. Although found throughout Queensland it is more common in the southern half of the State. A good food-fish. Grows to 40 centimetres and a weight of 2 kilograms.

Yellowfin or Common Bream *Acanthopagrus australis* PLATE 28:2

Colour silvery olive-green, silvery-white below; pectorals yellow, a black spot at base; ventrals and caudal yellowish. One of the best known food-fishes in Queensland. Commonly found everywhere and caught by both net and hand-line. Grows to at least 3 kilograms in weight.

Pikey Bream *Acanthopagrus berda* PLATE 29:1

Colour grey, dark silvery-grey, or dull olive-brown, darkest at the scale bases; white below; dorsal spines silvery; pectorals yellow; anal and ventrals blackish. Northern waters of Queensland. A good food-fish. Grows to 38 centimetres. Can be recognized by the stout second anal spine.

Family Sciaenidae: Teraglins and Jewfishes

Silver Teraglin *Otolithes ruber* PLATE 108:5

Colour silvery, darker on back, washed with iridescence of blue and pink above lateral line; a dark blotch on the gill-cover; paired fins and tail yellowish. A good food-fish which reaches a length of at least 81 centimetres. Flesh oily. Inhabits the northern waters of Queensland.

Teraglin *Atractoscion aequidens* PLATE 108:6

Colour bluish above, silvery on the sides and white below; inside mouth, lips, and edges of and inside the gill-covers bright yellow. A fine food-fish which is commonly captured in southern Queensland during the winter months. It grows to 1.3 metres and 9 kilograms in weight.

Jewfish or Mulloway *Argyrosomus hololepidotus* PLATE 109:1

Colour of young steel-blue above, becoming dark grey or brownish-green in adults; sides silver-grey, white below. Common in the waters of southern Queensland in the sea, bays and rivers at all times. Grows to 1.75 metres and at least 61 kilograms. Good food-fish up to about 60 centimetres.

Little Jewfish *Johneiops vogleri* PLATE 109:2

Colour silvery-grey above, pure silver on sides, whitish below; chest, belly and underparts of tail base ochre-yellow. This fish enters the rivers of southern Queensland in great numbers during winter months. Grows only to 30 centimetres but it is good eating. Known also as "River Perch".

Silver Jewfish *Nibea soldado* PLATE 109:3

Colour greyish or silvery, darkened with green or pink; white on sides and below; dorsals dusky, the rays darker, sometimes with a small brown spot anterior to each ray; ventrals white, other fins yellow. Common in North Queensland. Flesh of fair quality. Length to 76 centimetres.

Bearded Jewfish *Sciaena dussumieri* PLATE 29:2

Colour dark brown or greyish, sometimes nearly black, shot with golden reflections; silvery below, tinged yellow; fins reddish-brown; first dorsal stained with black; ventrals yellowish. Occurs in northern waters of Queensland but not commonly taken. Grows to 25 centimetres.

Family Carangidae

Northern Mackerel Scad *Decapterus russellii* PLATE 109:4

Colour dark blue or greenish-blue above; silvery below, with a diffused-yellow longitudinal lateral band; a narrow black spot on the gill-cover; fins hyaline or yellowish, tinged red; tail edged with yellow. Only occasionally found on the coast of Queensland. Grows to 45 centimetres.

Scad *Trachurus mccullochi* **PLATE 109:5**

Colour yellowish-green above, darkest on the head; sides golden, and silvery below, with reflections of bluish or pinkish iridescence; the fins are greenish-yellow; lobes of the tail yellow. A common fish in the bays and inlets of southern Queensland. It grows to 33 centimetres.

White Trevally *Caranx nobilis* **PLATE 110:1**

Colour silvery-yellow, tinged with blue on the back; a well-defined black spot on the gill-cover; fins yellow, excepting the tail, which is blackish at the tips of the lobes. Coastal and outside waters of southern Queensland. A good food-fish which grows to at least 1.1 metres.

Great Trevally *Caranx sexfasciatus* **PLATE 29:3**

Colour silvery, darker above; dorsal, anal, and tail dusky; pectorals yellow; young usually very golden-yellow, with broad dark vertical cross-bars. A large trevally commonly found along the entire coastline of Queensland. Grows to at least 1.25 metres and a weight of 36 kilograms.

Turrum *Carangoides emburyi* **PLATE 30:1**

Colour opalescent bluish above, light silvery below; a few yellow spots on the upper sides; fins olive; a small black blotch on the gill-cover. A fine game-fish and a terrific fighter when hooked. Average fish 10 to 13 kilograms. Known to reach 18 kilograms. Northern Queensland.

Herring Trevally *Alepes kalla* **PLATE 30:2**

Colour blue above, bronze on upper sides; silvery below, washed with gold; a large black shoulder spot; fins hyaline or yellowish. Found in the northern waters of Queensland and trawled to 25 fathoms. It is a wide-ranging species which grows to a length of about 30 centimetres.

Fringe-finned Trevally *Absalom radiatus* **PLATE 30:3**

Colour pale greenish-blue above and on sides, silvery-white below, with or without dark vertical bands; a large black gill-cover blotch; dorsal and anal membranes orange-yellow posteriorly; tail yellow, end of upper lobe black. Found in North Queensland waters. Length 33 centimetres.

Golden Trevally *Gnathanodon speciosus* PLATE 31:1

Young are canary-yellow on head, sides, and below, suffused with purple and blue above; alternately wide and narrow blackish cross-bands; fins yellow. Adults are silvery, washed with blue and without bands or spots. Common in North Queensland, uncommon south. Grows to 1 metre.

Cale-cale Trevally *Ulua mandibularis* PLATE 110:2

Colour olive-green, silvery below, sharply defined between; inner axil of pectoral black. Distinguished by the presence of long feather-like gill-rakers which extend into the mouth. It inhabits the northern waters of Queensland. Apparently not common. Length to 31 centimetres.

Diamond Trevally *Alectis indica* PLATE 110:3

Colour golden-bronze above, shading into silvery-white below. Young with broad dull blue bands from the back to below middle of the sides. Found along the entire coastline of Queensland but more common in the north. It is said to reach a length of 1.5 metres but such a size is rare.

Black Pomfret *Parastromateus niger* PLATE 31:2

Colour of body, fins, and tail greyish, greyish-brown, or brownish, tinged with blue; fins and tail sometimes edged with black. Commonly taken in the fish-traps in northern Queensland. It is an excellent food-fish and grows to 60 centimetres. It frequents the anchored lightships.

Large-mouthed Leather-skin *Scomberoides lysan* PLATE 32:1

Colour of head and back is golden-greenish-olive; rest of head and body golden-citrine or sulphur-yellow; a row of large round greyish indistinct spots along sides. Common in North Queensland southwards to the Capricorns. Grows to 1.25 metres and 9 kilograms weight. A poor food-fish.

Swallow-tail or Dart *Trachinotus russelli* PLATE 110:4

Colour of upper surfaces dark blue-grey; sides silvery, white below; dark oval spots on sides; dorsal, caudal, and anal lobes indigo-blue. Common in the surf and breakers along the coast of southern Queensland. Good food and sporting fish. Grows to 60 centimetres.

Runner *Elagatis bipinnulatus* PLATE 32:2

Colour very dark blue or greenish, whitish below; three pale yellow longitudinal bands from head to tail, between which are two blue ones; tail greenish or yellowish. A widely distributed open-sea fish found in all tropical and temperate seas. Grows to 1.25 metres. A good food-fish.

King Amber-jack *Seriola lalandi* PLATE 32:3

Colour pinkish-blue above, lower half silvery, the two colours being separated by a greenish-yellow band from eye to tail; fins edged yellow. Commonly taken by trolling-line in the southern waters of Queensland. Grows to 1.75 metres and a weight of 67 kilograms. Flesh only fair quality.

Pilot-fish *Naucrates ductor* PLATE 110:5,6

Colour bluish or blackish above, silvery on sides and lighter below, with vertical dark cross-bands. These fish associate with sharks in order to obtain scraps of food overlooked by their hosts. They are said to pilot sharks. Grow to 60 centimetres. Found in all warm seas.

Family Menidae: Moon-fishes

Moon-fish *Mene maculata* PLATE 33:1

Colour deep blue along back; sides and belly silvery; large oval spots of slaty-blue on upper part of body. This distinctively-shaped fish usually inhabits the deeper coastal waters of northern Queensland but has also been taken in the coastal fish-traps. It grows to 20 centimetres.

Family Mullidae: Surmullets or Goat-fishes

Mottled Goat-fish *Upeneus tragula* PLATE 111:1

Colour dark umber-brown or greyish-yellow; head and body mottled or speckled with black or brownish spots; a dark stripe from the eye to the tail-base; tail with oblique dusky bars. It is a very common fish along the coast of Queensland. Attains to a length of 35 centimetres. Its flesh is excellent.

Gold-striped Goat-fish *Mulloidichthys vanicolensis* PLATE 111:2

Colour scarlet, rose-pink, orange, yellow, or olive above, lighter below; a bright yellow (or sometimes red) stripe from head to tail; fins yellowish. Varies greatly in colour. An uncommon Queensland species. Grows to about 40 centimetres. Recognized by the lateral stripe.

Black-spot Goat-fish *Parupeneus signatus* PLATE 111:4

Colour red, upper surface clouded with olive-brown; a broad, obscure, dusky band from gill-covers to below rayed dorsal; a large black saddle-like spot over tail-base, frequently preceded by a pinkish blotch. Common in southern Queensland. Grows to 45 centimetres and 1.5 kilograms.

Family Cheilodactylidae: Morwongs

Morwong *Nemadactylus douglasii* PLATE 111:5

Colour purplish-grey, each scale with a greenish-golden, dark-edged spot; a golden band edged with purple iridescence from eye to snout; spinous dorsal yellow, also yellow spots on soft dorsal and anal fins. Frequents deeper outside waters of southern Queensland. Length 60 centimetres.

Magpie Morwong *Cheilodactylus gibbosus* PLATE 111:3

Colour silvery-white, with oblique brownish or blackish bands; the mouth is encircled with a dark brownish ring; edge of the gill-cover scarlet. This fish occasionally wanders into the waters of southern Queensland from New South Wales. It grows to a length of 30 centimetres.

Family Leiognathidae: Pony-fishes or Pouters

Pig-nosed Pony-fish *Secutor ruconius* PLATE 33:2

Colour bluish-silvery or brownish above, silvery below; a series of dark vertical lines or spots along side of back. Although not usually met with by casual observers this fish is commonly taken in northern waters of Queensland by the prawn-trawlers. Grows to 13 centimetres.

Black-banded Pony-fish *Equulites moretoniensis* PLATE 111:6

Colour iridescent pink, silvery-white below, with oblique wavy light blue bars and spots above; fins tinged yellow. Has been recorded from both ends of Queensland—the Gulf of Carpentaria and Moreton Bay, but so far not between these places. Grows to about 8 centimetres.

Family Gerridae: Silver-bellies

Thread-fin Silver-belly *Gerres punctatus* PLATE 112:1

Colour bluish-grey above, silvery below, with rows of oblong, bluish spots on upper half of body. The young have vertical bands. Commonly found in the waters of North Queensland. Occasionally stragglers have been captured as far south as Moreton Bay. It grows to about 25 centimetres.

Silver-belly *Gerres filamentosus* PLATE 112:2

Colour brilliant silvery, the back bright greenish or brownish; dorsal fins black-edged. A common fish which inhabits shallow sandy flats and bays in shoals, where it is taken by fishermen when netting mullet, bream, etc. Length 25 centimetres. Flesh of very sweet quality.

Family Chaetodontidae

Spotted Butter-fish *Scatophagus argus* PLATE 34:1

Colour bluish-grey, greenish, olive, or blackish, silvery or yellowish below, with smaller and larger, more or less numerous, round or oval brownish spots. Common on the Queensland coast. Stings from the spines of this fish are very severe. Grows to at least 31 centimetres.

Southern Butter-fish *Selenotoca multifasciata* PLATE 112:3

Colour of back and upper sides greenish, silvery below, with broad dark green vertical bands of elongate spots and horizontal stripes below on belly. Common in southern Queensland in estuaries and bays. It is a good food-fish but the flesh is soft. Grows to 40 centimetres.

Beaked Coral-fish *Chelmon rostratus* PLATE 34:2

Colour pearly-white, yellowish posteriorly; four orange-yellow vertical bands, the first three with dark brown margins, the last bordered with pale blue; a large eye-like spot on the dorsal. Common on the reefs of

Queensland from north to south. Grows to a length of about 20 centimetres.

Truncate Coral-fish *Chelmonops truncatus* PLATE 112:5
Colour white, all the spines yellow, with vertical bands of dark brown or black. Common in New South Wales but somewhat rare in the waters of southern Queensland, where it has been trawled down to 30 fathoms. Not yet recorded north of Double Island Point. Grows to 20 centimetres.

Long-bill *Forcipiger longirostris* PLATE 35:1
Colour brilliant yellow, orange posteriorly; a large, triangular black patch over nape, upper surface of head and side down to level of lower edge of eye; tail greyish. Has been recorded from Cape York and Moreton Bay but not yet between. Uncommon. Grows to 23 centimetres.

High-fin Coral-fish *Coradion altivelis* PLATE 112:6
Colour yellowish, with greyish vertical bands and a blackish band encircling the base of the tail. A little-known fish from the Wide Bay district of southern Queensland. Only two specimens are known, the largest of which measures 1.35 centimetres. Nothing is known regarding its habits.

Six-spined Butterfly-fish *Parachaetodon ocellatus* PLATE 35:2
Colour silvery, with yellow, black-edged vertical bands; a large black eye-like spot on the dorsal fin is included in the golden prolongation of the fourth band. A common fish in the waters of Queensland, usually among the many reefs and corals. Grows to a length of 18 centimetres.

Thread-fin Butterfly-fish *Chaetodon auriga* PLATE 36:1
Colour pearl-grey in front and below, tinged with violet, changing abruptly to deep orange-yellow behind and on vertical and caudal fins; a series of violet-grey lines running upward and backward with others running downward and backward. Common on coral reefs. Length 23 centimetres.

Criss-cross Butterfly-fish *Chaetodon vagabundus* PLATE 36:2
Colour varies from pale yellow to grey, with upward, oblique stripes

anteriorly, from which downward directed oblique stripes run posteriorly; vertical fins and tail yellow, with black stripes. A very common fish found on the northern coral reefs. Grows to about 20 centimetres.

Lined Butterfly-fish *Chaetodon lineolatus* PLATE 37:1

Colour silvery-grey, with narrow, black vertical lines down the scale-rows; a broad black band on head; another broad black band from middle of spinous dorsal, along base of soft dorsal, down across base of tail to base of anal. Found in the coral reefs. Grows to 30 centimetres.

Rainford's Butterfly-fish *Chaetodon rainfordi* PLATE 112:7

Colour canary-yellow, with darker orange cross-bands. This very pretty little butterfly-fish is common everywhere among the coral growths of northern Australia and southwards to about the vicinity of Noosa Heads, southern Queensland. It grows to about 14 centimetres only.

Golden-striped Butterfly-fish *Chaetodon aureofasciatus* PLATE 37:2

Colour purple-drab, passing into yellow below and orange-yellow on soft dorsal, caudal and anal; a narrow band through eye and another through pectoral base is bright yellow, bordered with black and violet. Common as far south as the Whitsunday Passage. Length 15 centimetres.

Right-angled Butterfly-fish *Chaetodon trifascialis* PLATE 38:1

Colour olive-brown, paler below; two large, elongate pale blotches on side, with numerous dark violet stripes, bent to a right-angle; a yellow-bordered black band runs through eye to throat. Although it has been recorded from North Queensland it is rarely seen. Length 15 centimetres.

Family Scorpididae

Stripey *Microcanthus strigatus* PLATE 112:4

Colour is silvery-white, with broad black bands sloping slightly downwards posteriorly from the horizontal. The Stripey is a very common little fish which frequents the rock-pools along the coast of southern Queensland. It grows to a length of about 15 centimetres.

Family Chaetodontidae

Feather-fin Bull-fish *Heniochus acuminatus* PLATE 38:2

Colour pearly-white or yellowish-white; body with two wide cross-bands of blackish-brown, the first including the ventrals, the second one covering the posterior half of the anal. A handsome fish occasionally met with in the northern waters of Queensland. It grows to 25 centimetres.

Family Pomacanthidae: Angel Fishes

Scribbled Angel-fish *Chaetodontoplus duboulayi* PLATE 39:1

Colour of body purplish-brown or black, with or without many narrow wavy pale lines; yellow of the head and breast extends along the base of the dorsal fins to the tail. Mainly inhabits the northern waters of Queensland. It is uncommon south of the Capricorns. Length 28 centimetres.

Emperor Angel-fish *Pomacanthus imperator* PLATE 113:1

Colour golden-brown, with numerous narrow yellow or orange bands more circular in the young and more numerous and semi-horizontal in the adult. A beautiful fish found on the coral reefs of northern Queensland. It grows to 38 centimetres. Flesh of good edible quality.

Zebra Angel-fish *Pomacanthus semicirculatus* PLATE 113:2

Colour of the young is very dark indigo with transverse white bands alternately narrow and broad; the bands disappear in the adults, which are yellowish and greyish with blackish or blue spots. Inhabits the coral reefs of Queensland from the Capricorns northward. To 38 centimetres.

Six-banded Angel-fish *Euxiphipops sexstriatus* PLATE 114:1

Colour yellowish-brown or golden-green, each scale with a dark blue centre; vertical bands of dark blue across the body; a broad band of pearly-white across the head. A very common fish on the coral reefs of Queensland. Grows to a length of 50 centimetres and is a good food-fish.

Black and Gold Angel-fish *Centropyge bicolor* PLATE 114:2

Colour of the anterior half of the body is orange or lemon-yellow, posterior half violet-black or deep black; a broad blue-black saddle above the eyes.

This small but beautiful fish dwells among the coral growths on the reefs of Queensland. It grows only to about 14 centimetres.

Family Ephippidae

Sickle-fish *Drepane punctata* PLATE 39:2

Colour bright silvery, with purplish reflections; usually a vertical series of blackish spots on the upper sides. A common fish along the coast of Queensland southwards to about Gladstone. Frequents the sea and brackish estuaries. Flesh is somewhat dry. Length to 45 centimetres.

Batfish *Platax pinnatus* PLATES 40, 41

Colour blackish-brown, tinged with purple; young are yellowish or reddish-brown with three diffused black bands and with the anterior dorsal and anal rays greatly prolonged. A common fish found along the coast of Queensland. Said to grow to 76 centimetres and a weight of 22 kilograms or so.

Family Enoplosidae: Old Wife

Old Wife *Enoplosus armatus* PLATE 113:3

Colour silvery-white, with blackish or dark brown vertical bands; the posterior membrane of the dorsal and anal rays is bright red; tail pink. A southern species which wanders into southern Queensland waters. It grows to 23 centimetres. Though small, its flesh is very sweet eating.

Family Pomacentridae: Anemone- or Damsel-fishes, Pullers

Spine-cheeked Anemone-fish *Premnas biaculeatus* PLATE 113:4

Colour dark or light cherry-red to brownish-red all over body and fins; three transverse pearl-white bands, narrowly edged with black (as in illustration); fins narrowly edged black. Inhabits anemones on the reefs of North Queensland. Males to 11 centimetres and females to 16 centimetres.

Orange Anemone-fish *Amphiprion percula* PLATE 42:1

General colour, including fins, orange or orange-brownish; three transverse bluish-white bands (as in illustration) are usually narrowly edged black. Subject to much colour variation. Associated with sea-anemones. Very common. Males to 8 centimetres, the females larger.

Green Puller *Mecaenichthys immaculatus* PLATE 114:6

Colour pale olive-green, silvery-white below, the sides washed with bronze; vertical fins tinged with violet and narrowly edged with blue. A southern fish which has been recorded from Cape Moreton, southern Queensland. A rather uncommon species which grows to a length of 15 centimetres.

Blue Puller *Chromis caeruleus* PLATE 42:2

Colour blue, greenish-blue, or green, whitish below; each scale of head and sides with a blue spot. In some examples there is a bright golden-yellow spot on each scale and the fins are orange. A very common fish on the coral reefs of Queensland. Length to about 13 centimetres.

Banded Humbug *Dascyllus aruanus* PLATE 114:7

Colour is beautiful pearly-white, with three broad, black cross-bands; tail white. A handsome little fish commonly seen in hundreds among the coral growths on the numerous reefs of Queensland. It grows to a length of 8 centimetres. Named "Humbug" after the popular confection.

Spotty-tail *Acanthochromis polyacanthus* PLATE 114:5

Colour dark purplish-brown as far back as a line from the origin of the soft dorsal to the anal spines; posterior to this line is greyish; sometimes with numerous black spots on body as well as fins; anus in a black area. Reefs of North Queensland. Length to 18 centimetres.

Jerdon's Damsel-fish *Pristotis jerdoni* PLATE 42:3

Colour olive or yellow; head with rows of light blue spots; a row of light blue lines along centres of scales; fins dark-coloured; a blackish spot at pectoral base. Has been trawled along the coast of Queensland at depths ranging from 11 to 30 fathoms. Length to 15 centimetres.

Orange-tipped Damsel-fish *Stegastes apicalis* PLATE 115:2

General colour blackish-brown, including fins and tail; outer one-third of spinous dorsal bright orange, through which runs a violet submarginal band; tip of upper caudal lobe bright orange. A somewhat uncommon fish found on the reefs of North Queensland. Length 10 centimetres.

Yellow Damsel-fish *Pomacentrus moluccensis* PLATE 115:1

General colour cadmium-yellow, passing into strong orange colour on the tail, soft dorsal and anal fins; a small, pale blue or black spot at the commencement of the lateral line and a black one in the axil of the pectoral. Common among coral growths in Queensland. Length 7.5 centimetres.

Palmer's Damsel-fish *Abudefduf bengalensis* PLATE 115:3

Colour greenish-grey, with seven black vertical cross-bands; head and throat like the back, with scattered silvery spots and bars; dorsal, anal, and ventral fins blackish. A common fish along the Queensland coast, particularly in the south. It grows to at least 18 centimetres.

Clouded Damsel-fish *Amblyglyphidodon curacao* PLATE 115:4

Bright pale green above, silvery below, with dusky bands about three scales wide, narrowing and becoming indistinct below; each scale within the dark bands pale-edged. Recognized by the indistinct bands and the circular body. Found on the reefs of Queensland. Grows to 13 centimetres.

Orange-tailed Damsel-fish *Neopomacentrus nemurus* PLATE 115:5

Colour brownish-black, passing into grey below; posterior portions of soft dorsal, anal, and whole of the tail cadmium-yellow and terminating in long black rays. Very commonly seen playing in dozens among the corals on the reefs of North Queensland. Grows to 7.5 centimetres.

Large-scale Parma *Parma oligolepis* PLATE 115:7

Colour is uniform brownish; the borders of the fins and the margins of the scales of the body are somewhat darker. This deep-bodied fish with large scales, is found in the coastal waters of the southern parts of Queensland. It grows to a length of at least 22 centimetres.

Big-lip Damsel-fish *Cheiloprion labiatus* PLATE 115:6

Colour is uniform brownish, the fins somewhat darker, the ventrals blackish; lips pale in colour, very wide and with fleshy projections. A little fish, sombre-coloured, recognized by its swollen and curled lips. It is found on the reefs of North Queensland. Grows to 8.5 centimetres.

Family Labridae: Wrasses, Rainbow-fishes, Pig-fishes, Tusk-fishes

Giant Pig-fish *Achoerodus unicolor* PLATE 116:1

Colour of male is purplish-brown, sides and below yellowish-brown, each scale with an orange spot or bar; orange around eye and on snout; adult more uniform. The female is uniform dull brick-red. This large fish is found in southern Queensland. It grows to 1.25 metres and 18 kilograms.

Harlequin Tusk-fish *Lienardella fasciatus* PLATE 43:1

Colour is blue-black posteriorly, pale green on the sides, breast, and belly; the head and body are crossed with vertical scarlet bands; tail scarlet or pink. This most beautiful fish commonly frequents the coral reefs of North Queensland. It grows to about 25 centimetres.

Purple Tusk-fish *Choerodon cephalotes* PLATE 43:2

Colour of forehead and snout purplish-green, with narrow transverse orange bands; throat yellow, with blue bars; body with blue vertical lines; tail with orange and blue vertical lines. Common along the whole Queensland coastline. Grows to at least 35 centimetres, probably much more.

Venus Tusk-fish *Choerodon venustus* PLATE 44:1

Colour of upper part of head green, becoming bluish towards snout; chin sky-blue; body olive-brown above, rose-colour below; scales with a blue spot; fins and tail blue and gold. Common on the off-shore reefs and rocky shores of Queensland. Grows to 1.25 metres. A good food-fish.

Blue Tusk-fish *Choerodon albigena* PLATE 44:2

Colour bluish-green, greyish on head; snout with lines of dull orange and bluish-green; chin and throat pink; tail dark purplish-blue; dorsal and anal fins orange, with blue spots and lines. Common on the reefs of Queensland. Grows to 71 centimetres and 8 kilograms. Good food-fish.

Golden-spot Pig-fish *Bodianus perditio* **PLATE 116:2**

Colour of the head and body is bright red, suffused with dark orange, then violet posteriorly; yellow spots on the head and forepart of body; a bright yellow patch on the back, bordered black posteriorly. A fairly common fish on the reefs of Queensland. To 53 centimetres. A good food-fish.

Sling-jaw *Epibulus insidiator* **PLATE 46:1**

Colour dark brown or olive-green; before eyes and cheeks grape-green; tail very dark reddish-olive; body-scales edged black or dark red. Very variable; some forms are light yellow. Recognized by its extraordinary jaws. Found on the reefs of North Queensland. Grows to about 33 centimetres.

Cockatiel-fish *Pteragogus flagellifera* **PLATE 116:3**

Colour of the male is usually a deep blue, marked with gold. The female may be brick-red with blue lines on the head and blue-green spots on the head and body. Most variable in colour. A small fish, mainly taken by prawn-trawlers in Queensland. Grows to about 20 centimetres.

Tube-lip *Labrichthys unilineata* **PLATE 116:4**

Colour very dark brown, with obscure stripes of bronze and blue; head with navy-blue stripes; lips yellow; dorsal and anal fins edged blue; a blue, curved streak across tail. Lips form a tube when mouth is closed. Reefs of North Queensland. Grows to a length of 15 centimetres.

Blue Streak *Labroides dimidiatus* **PLATE 116:5**

Colour brilliant blue; a black longitudinal band from snout through eye to end of tail, narrow in front but becoming broader posteriorly so as to occupy most of the tail, except upper and lower rays which are blue. A very active fish among the corals of Queensland. 10 centimetres.

Günther's Rainbow-fish *Pseudolabrus guntheri* **PLATE 116:7**

Colour chocolate-brown, which breaks into broad stripes on sides and these extend on to belly and become reddish, with apple-green between;

head apple-green with red stripes around eye. Vertical bars more distinct in young. Inhabits southern Queensland. Grows to 18 centimetres.

Lined Rainbow-fish *Stethojulis strigiventer* PLATE 116:6

Colour of the upper half of head and body from snout to tail-base is sage-green; lower half of head and body with pale olive lines longitudinally, becoming silvery on the belly; tail olive-green with bluish-white bars. Reefs of North Queensland. It grows to 13 centimetres.

Black-eyed Thick-lip *Hemigymnus melapterus* PLATE 45:2,3

Colour extremely variable; up to 13 centimetres anterior half of body is light-coloured, the posterior half dark-coloured, the two colours sharply defined. In adults the light and dark colours disappear (as shown in the plate). Coral reefs of Queensland. It grows to 90 centimetres.

Clouded Rainbow-fish *Halichoeres nebulosus* PLATE 46:1

Colour greenish or yellowish with numerous violet or reddish curved, irregular markings on head and dorsal fin, the latter having a black spot on the anterior dorsal rays. A handsome fish, very variable in colour, living among corals on the Queensland coast. Length 10 centimetres.

Saddled Rainbow-fish *Halichoeres margaritaceus* PLATE 46:2

Colour is bright blue above, with a network of dark reddish-brown bordering the scales and extending almost to a line with the pectoral base but exceeding this line towards the tail; a pink, curved stripe below eye. Common on the coral reefs of Queensland. Grows to 10 centimetres.

Maori *Ophthalmolepis lineolatus* PLATE 117:1

Colour of back is red, with many pale blue spots; sides with a broad pink longitudinal band anteriorly and a dark band posteriorly, which are usually separated by a narrow light band; belly yellowish; head with pale blue stripes. Inhabits southern Queensland. To 40 centimetres.

Comb-fish *Coris picta* PLATE 117:2

Colour of upper snout greyish, followed by a wide red stripe along the back to tail-base; below this, on the snout, is a dark chocolate band from eye to tail, its lower edge becoming comb-like; lower half of body lemon-yellow.

Common in southern Queensland. Grows to 28 centimetres.

Ringed Wrasse *Hologymnosus semidiscus* **PLATE 117:3**

Colour rosy, yellowish below, with numerous narrow yellow or brown, or blue edged with brown, vertical cross-bars; yellowish or bluish streaks on head; dorsal fin rosy with basal white spots. A handsome though very variable fish not commonly met with. Grows to 50 centimetres.

Moon Wrasse *Thalassoma lunare* **PLATE 47:1**

Head is rich violet, with arched bands of green; dark greenish-blue body, each scale with vertical red streak; dorsal and anal fins broadly edged yellow. This active and handsome fish is very variable in colour. It frequents the coral reefs of Queensland and grows to about 25 centimetres.

Sharp-nosed Wrasse *Cheilio inermis* **PLATE 117:4**

Colour dark ivy-green, blue below; the lighter edges of the scales form a lattice pattern; head olive-green above, sky-blue below, divided by a narrow orange line. Subject to great colour variation. Fairly common in southern Queensland, rarer northwards. Grows to 50 centimetres.

Scribbled Wrasse *Anampses geographicus* **PLATE 117:5**

Colour olive, with purplish and blue tones; vertical bar of scales bright blue with darker edges, changing on belly to horizontal stripes; head with fine lines; greenish above pectorals. A distinctively marked little fish found on the reefs of North Queensland. Grows to 15 centimetres.

Bar-cheeked Wrasse *Hemipteronotus taeniourus* **PLATE 117:6**

Colour of body, with exception of belly, is dark olive-green, each scale with a light green spot and edged dark green; belly bright russet, each scale edged white; head light green with blackish bars. Identified by the "brick wall" effect on the belly. North Queensland. 25 centimetres.

Scarlet-breasted Maori Wrasse *Cheilinus fasciatus* **PLATE 47:2**

Colour of head dark green, with scarlet bars radiating from eye; a broad scarlet patch from origin of spinous dorsal down to breast and anterior part of belly; rest of body pinkish-red with blackish vertical bands. Inhabits the coral reefs of North Queensland. Length 35 centimetres.

Hump-headed Maori Wrasse *Cheilinus undulatus* PLATE 47:3

Top of head and along back dark purple, fading into green on upper sides, then lighter green, into yellowish-green below; each scale with a broad vertical bar of dark violet and orange lines. The largest species of the family. Inhabits the reefs of Queensland. Length 2 metres.

Keel-head *Hemipteronotus jacksonensis* PLATE 47:7

Colour opalescent salmon, with oblique blue hair-lines above; two or three rows of scales bordering the ventral fins back to tail-base, each with a large blue basal blotch; fins orange; tail barred. A fish which straggles into southern Queensland from the south. Grows to 19 centimetres.

Family Odacidae

Herring Cale *Olisthops cyanomelas* PLATE 47:8

Female: head and body olive-brown above, orange on sides and salmon on belly; head with dark blue lines and spots; body scales with large blue spot. Body of male is uniform bluish-black, lighter below; tail lobes with a blue band near margins. Southern Queensland. Size 45 centimetres.

Family Scaridae: Parrot-fishes

Marbled Parrot-fish *Leptoscarus vaigiensis* PLATE 118:1

Colour olive-brown above, mottled dark brown; sides olive-green, mottled brick-red; pale greenish-cream below, with yellowish; flecks of brick-red and pink on cheeks; tail olive-yellow marbled with darker olive. Not commonly met with in Queensland. Grows to about 40 centimetres.

Half-toothed Parrot-fish *Calotomus spinidens* PLATE 118:2

Colour olive-brown above, finely dotted with darker brown; sides pale green, whitish below; several incomplete fawn-coloured brownish-ringed spots on back and sides; dorsal and anal fins with mottlings of pale reddish-brown or brick-red. Southern Queensland. Length 29 centimetres.

Surf Parrot-fish *Scarus fasciatus* PLATE 48:1,2

Male is light bright green, passing into coral-red or pinkish-red below, each scale with a basal blotch of coral-red; tail orange and green. Female is turquoise-blue, each scale edged with bright yellow. Seen in schools over the coral reef. Good food-fish. Length 50 centimetres.

Family Ammodytidae: Sand Lances or Sand Eels

Sand Lance *Bleekeria vaga* PLATE 118:4

Colour in preservative is uniform sandy-yellow, the gill-covers blackish. Probably silvery in life. Originally described from Lord Howe Island and since taken in New South Wales, this fish has also been taken in southern Queensland. It grows to at least 16 centimetres in length.

Family Opisthognathidae: Smilers or Jaw-fishes

Harlequin Smiler *Merogymnus eximius* PLATE 118:6

Colour golden or golden-brown above; sides with large round or oval golden spots, separated by broad blue interlacing bands; belly and end of tail violet. A beautiful fish commonly found in the southern waters of Queensland as far north as the Capricorns. Grows to 36 centimetres.

Speckled Pug *Opistognathus papuensis* PLATE 48:3

Colour is pale reddish-brown, everywhere spotted with dark blue or black, the spots being largest on the upper part of the tail and trunk, smallest on the head. Occasionally captured in the waters of North Queensland. It is said to be a good edible fish and grows to 43 centimetres.

Family Mugiloididae: Weevers

Bar-faced Weever *Parapercis nebulosus* PLATE 49:1

Colour bright reddish-pink above, the scales with darker edges; red saddle-like bands across back and upper sides; blue lines on snout. Commonly taken in southern Queensland but not often met with in North Queensland. Its pungent spines can inflict painful wounds. Grows to 30 centimetres.

Family Uranoscopidae: Star-gazers

Star-gazer *Ichthyscopus sannio* PLATE 49:2

Upper part of the head, along the back and upper sides brownish, enclosing pure white, round or oval spots which are largest on the back and sides; tail with yellowish streaks. Not commonly met with because of its habit of burying itself in the sand. Grows to at least 63 centimetres.

Family Acanthuridae: Surgeon-fishes, Unicorn-fishes

Bristle-toothed Surgeon-fish *Ctenochaetus strigosus* PLATE 118:3

Colour reddish-brown or chocolate-brown, with numerous longitudinal wavy narrow blue lines along sides; head and breast with numerous small brownish-red or golden spots. A common species on the coral reefs of the South Seas but rarer in Queensland. It grows to about 28 centimetres.

Flag-tail Surgeon-fish *Paracanthurus hepatus* PLATE 118:7

Colour brilliant ultramarine, becoming indigo at base of pectoral and on basal half of dorsal and anal membranes; a black band from eye to tail encloses a "blue island" above the pectoral. A beautiful fish found in the waters of North Queensland. It grows to about 25 centimetres.

Five-banded Surgeon-fish *Acanthurus triostegus* PLATE 49:3

Colour pale greenish or light olive, sometimes yellowish; chin, belly, throat, and a narrow stripe along base of anal, white; a black line along middle of snout; sides with narrow black bands. Large schools frequent pools in the reefs of North Queensland. Grows to 25 centimetres.

Orange-spot Surgeon-fish *Acanthurus olivaceus* PLATE 118:5

Colour of the body and head dark olive-brown or olivaceous, with a broad, deep yellow or orange longitudinal bar on the gill-covers and above the pectorals, margined with black or brown; a milk-white crescent on margin of the tail. Found on Queensland reefs. Grows to 40 centimetres.

Sail-finned Surgeon-fish *Zebrasoma veliferum* PLATE 118:8

Colour dark brown or almost black; sides of the snout and chin with yellow spots; broad golden-yellow, lemon-yellow, or white bands on the sides. This very handsome fish exhibits great variation of colour. It inhabits the coral reefs of North Queensland and grows to 40 centimetres.

Brown Unicorn-fish *Naso unicornis* PLATE 50:1

Colour is dark olive, brownish, or greyish-black; dorsal and anal fins are brown or light orange, with dark blue edges and with bluish spots and lines. It is very variable in colour. Found on Queensland reefs. The horn lengthens with age. Grows to 58 centimetres.

Family Zanclidae: Moorish Idols

Moorish Idol *Zanclus cornutus* PLATE 119:2

Colour canary-yellow shading into white, with two wide dark brown or black cross-bands, edged with blue or pearly-white stripes. A very beautiful and graceful fish when seen gliding among the coral growths. More common in the northern waters of Queensland. Grows to 18 centimetres.

Family Siganidae: Spinefeet or Rabbit-fishes

Fox-face *Siganus vulpinus* PLATE 119:1

Colour of the upper part of the body, the dorsal, anal, and tail orange, rest of the trunk clear yellow; a blackish-brown band on top of the head, behind which is a wide diagonal pearly-white band with reddish dots; breast black. Reefs of North Queensland. Length 23 centimetres.

Black Spinefoot *Siganus rivulatus* PLATE 119:3

Colour olive or chocolate-brown, with lighter blotches; back and sides with numerous small close-set light blue spots, also much larger scattered dark brown or blackish spots; a black shoulder spot usually present. Colour variable. Common on Queensland coast. Length 35 centimetres.

Golden-lined Spinefoot *Siganus lineatus* PLATE 50:2

Colour slaty-blue with longitudinal yellow lines and spots; face and cheeks yellow, with bright blue lines; a large dusky yellow spot below last dorsal rays. A common Queensland fish especially plentiful on the reefs of North Queensland. It grows to a length of 38 centimetres.

Family Trichiuridae: Cutlass-fishes or Hair-tails

Cutlass-fish *Lepturacanthus savala* PLATE 119:4

Colour greyish along back, silvery on sides and below; fins pale yellow, the upper half of dorsal dusky with blackish dots. A remarkable fish which inhabits the open seas, bays and estuaries of Queensland. Mouth with long, strongly compressed teeth. Length to 1.35 metres.

Family Istiophoridae: Sail-fishes, Spear-fishes, Marlins

Indo-Pacific Sail-fish *Istiophorus platypterus* PLATE 50:3

Colour slaty to purplish-blue, passing into whitish below; body with light grey cross-bands which break into variable spots with age; dorsal fin dark prussian-blue with blackish spots. A remarkable fish often taken by mackerel fishermen on the Queensland coast. Grows to 3.5 metres.

Black Marlin *Makaira indica* PLATE 119:5

Colour uniform dark grey or bluish-black on back, passing into almost white below; fins blue-blackish, except the anal which is dark grey. The Black Marlin inhabits outside waters of southern Queensland. Reaches a weight of over 373 kilograms and length of 4.25 metres.

Family Scombridae: Mackerels and Tunas or Tunnies

Long-jawed Mackerel *Rastrelliger kanagurta* PLATE 51:1

Colour bluish above, with greenish tinge anteriorly; a row of greyish spots on each side of spinous dorsal and usually one on base of soft dorsal origin; longitudinal lines above lateral line. Visits the waters of North Queensland in winter months. Grows to 30 centimetres.

Common Mackerel *Scomber australasicus* PLATE 51:2

Colour of back and upper sides bright green with wavy transverse bars or streaks of dark green; sides with darker spots; pearly-white below with iridescence. Swims in great shoals in southern Queensland waters. The flesh is of good edible quality. Grows to about 50 centimetres.

Big-scaled Mackerel *Gasterochisma melampus* PLATE 120:1,2

Colour bluish-black above, silvery-white below. Undergoes great changes with growth, especially the ventral fins. Up to size of 38 centimetres these fins are very large, becoming smaller with age. Rare in Queensland waters. An open-sea fish which grows to a length of 1.75 metres.

Scaleless Tuna *Gymnosarda unicolor* PLATE 51:3

Colour dark bluish to violaceous, greyish-white below; tip of soft dorsal and anal clear; dorsal finlets bluish, anal finlets yellowish. Inhabits outside waters of North Queensland. Said to grow to 2.5 metres but Queensland fish are about 1.5 metres. Flesh of good flavour.

Skipjack Tuna *Katsuwonus pelamis* PLATE 120:3

Colour deep bluish-black above; sides silvery with dark longitudinal bands. Most commonly met with in great shoals in the southern waters of Australia but is only occasionally taken in southern Queensland. It grows to a length of about 1 metre and is a valuable canning fish.

Mackerel Tuna or Little Tuna *Euthynnus affinis* PLATE 51:4

Colour dark bluish or green above, passing into silver; white below; greenish-blue wavy oblique streaks above lateral line; fins dusky; a few blackish spots on breast. Found in shoals on the Queensland coast. The flesh is dark but of fair canning quality. Grows to about 1.25 metres.

Watson's Bonito *Cybiosarda elegans* PLATE 52:1

Colour of head sea-green to level of snout, below which is yellowish-white; back sea-green, passing into greyish-green then white below; many small black spots on the back and longitudinal black stripes below. Common on Queensland coast. Length 40 centimetres. Good food-fish.

Pacific Yellow-finned Tuna *Thunnus albacares* PLATE 52:2

Colour deep prussian-blue above, sides yellow, whitish below; the dorsals, pectorals and ventrals yellow; anal and dorsal and anal finlets brilliant yellow. Roams Queensland waters in shoals during the spring and summer months. Weight of Queensland fish is about 12 kilograms.

Northern Blue-finned Tuna *Thunnus tonngol* PLATE 52:3

Colour greyish-blue above, sides silvery-greyish, the colours being separated by a blue stripe; rows of colourless spots usually present; dorsal and anal finlets yellow. Inhabits Queensland coastal waters in shoals. The pale flesh is soft but of good flavour. Length to 1.25 metres.

Australian Bonito *Sarda australis* PLATE 52:4

Colour blackish-green above, silvery on sides and below; dark longitudinal lines on sides. A common fish on the Queensland coast, being found in great shoals. Flesh is light in colour and of good flavour and canning quality. Grows to 1 metre but averages 50 centimetres.

PLATE 121:1,2

Narrow-barred Mackerel *Scomberomorus commerson*

Back and upper sides mottled iridescent blue and green, with purple and bronze colours; sides with transverse vertical bars of dark grey. One of the most important food-fishes of Queensland. The record fish is said to have weighed 58.75 kilograms and was 2.15 metres in length.

PLATE 121:3,4

Queensland School Mackerel *Scomberomorus queenslandicus*

Head and upper part of back iridescent-green; cheeks and belly are silvery-white; sides with indistinct blotches; spinous dorsal fin black, with a large area of intense white. Inhabits bays and coastal waters of Queensland. Grows to 76 centimetres; averages about 1 to 2 kilograms.

Spotted Mackerel *Scomberomorus niphonius* PLATE 122:1

Colour bluish above, sides silvery-grey with dark grey spots. It frequents shallow reef waters on the coast of Queensland. Fished in northern waters from October onwards and in the south from December to April or May. Attains over 1 metre in length. A valuable food-fish.

Broad-barred Mackerel *Scomberomorus semifasciatus* PLATE 122:2

Colour of back is greenish-blue, with broad vertical bands of dark grey; with age the bands break up into spots or more or less fade out, and these fish become known as "Grey Mackerel". They inhabit the coastal waters of Queensland. Grows to at least 1 metre. Valuable food-fish.

Wahoo or Peto *Acanthocybium solandri* PLATE 122:4

Colour prussian-blue on back, silvery on sides and below, the two colours separated by a broad sea-green stripe; sides of body with silvery-green vertical bands. Lover of the open seas; not really common anywhere. Grows to at least 1.75 metres and 56 kilograms. A good food-fish.

Large-scaled Tunny *Grammatorycnus bicarinatus* PLATE 53:1

Colour of the back is brilliant metallic-green with tints of bright blue; upper sides yellowish-bronze-green; area between the two lateral lines silvery. Commonly taken with the mackerel over the many reefs along the Queensland coast. A good edible species. Grows to 1 metre.

Family Gobiidae: Gobies and Gudgeons

Mud-hopper *Periophthalmus regius* PLATE 123:1

Colour greyish-brown, with dark bars on the sides; a broad, black, white-edged, submarginal band on both dorsal fins. Commonly seen jumping about on the mud at low-tide on the Queensland coast as far south as Bundaberg; rarely seen south of that area. Length 14 centimetres.

Redhead *Paragobiodon echinocephalus* PLATE 123:2

Colour varies greatly. Often the body and fins are blackish, with the head flesh-coloured or red. Others are golden-yellow or green, with the head brownish, though sometimes green. A very common little fish amongst the coral growths on the Great Barrier Reef. Grows to 3.5 centimetres.

Oyster Goby *Arenigobius leftwichi* PLATE 123:3

Colour pale yellowish-brown; scales (except along ventral profile) edged dark brown; a row of dark blotches along middle of side. Abounds in pools left by the receding tide on the oyster-banks of the coast of Queensland. It hides and breeds among the oysters. Length to 7.5 centimetres.

Bynoe's Goby *Amblygobius bynoensis* PLATE 53:2

Colour greenish; back with brownish-violet cross-bands; sides with two blue-edged brownish-violet longitudinal bands; tail-base with a brown spot; neck with a series of brown, blue-edged spots. A handsome goby found in the waters of North Queensland. It grows to 9.5 centimetres.

Crested Mud Goby *Cryptocentroides cristatus* PLATE 123:4

Colour purplish or reddish-brown above, fading into lilac below; sides with alternate darker and lighter cross-bands; upper surface and sides of head and nape with many blackish spots. Lives in muddy areas in Queensland waters where it occupies deserted burrows. To 10 centimetres.

Cobra-fish *Parioglossus rainfordi* PLATE 123:8

Colour light green; a dark bluish-black mark on the tail-base; a violet-brown band extends along the middle of the side; tail with two broad oblique bars, the median rays pink. A little fish in North Queensland. Found living in logs drilled by "Cobra". Grows to 5 centimetres.

Lined Gudgeon *Valenciennea muralis* PLATE 123:9

Colour is pale green with pale pink, blue-edged longitudinal bands; the dorsal fin is pale green, with several pink wavy bands; a large black spot behind the third spine at its tip. A very pretty little fish found on the reefs of North Queensland. It grows to 11 centimetres.

Family Eleotridae

Mud Gudgeon *Ophiocara aporos* PLATE 123:5,6

Colour dark above, orange below; scales of the back and sides with dark spots; fin membranes dark, the rays orange; dorsal, ventrals, and anal fins edged red; tail spotted yellow. A large gudgeon found in the sea and the rivers of North Queensland. Grows to a weight of 68 grams.

Family Callionymidae: Dragonets

Fingered Dragonet *Dactylopus dactylopus* PLATES 53:3; 123:7

Colour of male: see colour-plate. In the female the membranes of the spinous dorsal are golden-brown with darker marblings and blue spots and bands; soft dorsal with dark brown blue-edged spots between the rays. Found along the whole coast of Queensland. It grows to 20 centimetres.

Rough-headed Dragonet *Callionymus limiceps* PLATE 123:10-13

Colour brown, closely speckled with brown spots, which form short lines and extend obliquely forward below the eye and are enlarged along the sides of the body; first dorsal with a large black spot. Commonly taken in the prawn-trawls in southern Queensland. To 25 centimetres.

Mandarin-fish *Synchiropus splendidus* PLATE 54:1

Colour very elaborate, as seen in the colour-plate. This strikingly handsome fish dwells among the living corals on the Great Barrier Reef. In life it is one of our most beautiful fishes. Known in Queensland from only two examples. It grows to a length of at least 8 centimetres.

Family Blenniidae: Blennies

Oyster Blenny *Omobranchus anolius* PLATE 124:1

Colour of head and body chestnut-brown or olive-green; body with many indistinct angulated transverse bars; head with silvery streaks and a dusky cheek-spot; tail orange. Lives and breeds in dead shells of oysters in southern Queensland waters. Grows to 7.5 centimetres.

Collared Blenny *Cirripectes variolosus* PLATE 124:2

Colour almost uniform dark brown or blackish, with greyish spots as large as the eye, forming wavy lines; dark blue vertical bars on the body; fringe of tentacles on the nape is black. A fish from the reefs of North Queensland and Torres Strait. It grows to 10 centimetres.

Lineated Blenny *Salarias fasciatus* PLATE 124:3

Colour brownish or yellowish-brown, irregularly variegated, spotted, and dotted with yellow and blue; region above gill-covers and along back thickly sprinkled with black "dots and dashes", which become thin wavy lines on sides. Common on reefs of North Queensland. Grows to 15 centimetres.

Many-toothed Blenny *Ecsenius mandibularis* PLATE 124:4

Colour light brown in formalin, with bluish tinge on sides; two rows of brown spots, upper row near back, the lower along middle of sides. The female differs from the male in the form of the fins and in lacking the occipital crest. It inhabits the Capricorns. Grows to 6 centimetres.

Hair-tailed Blenny *Xiphasia setifer* PLATE 124:5

Colour bright yellow, with dusky bands on dorsal and body; dorsal and anal broadly blackish-edged; caudal filament black. A rare open-sea fish known in Queensland from a single example taken in the surf at Stradbroke Island, southern Queensland. It grows to 50 centimetres.

Long-nosed Weed-fish *Heteroclinus nasutus* PLATE 124:7

Colour may be anything from dark green without markings to pale green with darker bands and silvery or transparent markings; dorsal is hyaline, tinged yellowish and with dark olive-green blotches. Lives in seaweeds in southern Queensland to Caloundra. It grows to 6 centimetres.

Family Notograptidae

Spotted Eel Blenny *Notograptus guttatus* PLATE 124:6

Colour reddish or brownish, with many blue spots on upper and lateral parts of body; dorsal with numerous less distinct blue spots. It frequents the waters of North Queensland as far south as the region of the Whitsunday Passage. Grows to 13 centimetres. Not commonly seen.

Family Peronedysidae

Snake Blenny *Peronedys anguillaris* PLATE 124:8,9

Colour in preservative is a uniform reddish-brown, the occiput and the gill-covers with a yellowish tinge. A small and apparently rare little fish so far known only from Moreton Bay in southern Queensland. It grows to about 8 centimetres and nothing is known of its habits.

Family Tripterygiidae

Jumping Joey *Lepidoblennius haplodactylus* PLATE 124:12

Colour (in preservative) yellowish, with a number of rather faint reddish-brown spots and double vertical bands along the body; fins of a more dusky-yellow; anal blackish-margined. Found uncommonly in the rock-pools of southern Queensland. It grows to about 13 centimetres long.

Family Congrogadidae

Eel Blenny *Blennodesmus scapularis* PLATE 124:10

Colour of body brownish, marbled darker; sides of head with small yellowish spots; a black, yellow-edged eye-like spot in scapular region and an undulated yellowish line along middle of nape and head. A small fish very occasionally found in coral pools of Queensland. Grows to 8 centimetres.

Mud Blenny *Congrogadus subducens* PLATE 124:11

Colour generally olive, covered with rust-red blotches, those on belly, chin, and throat alternating with spots of bluish-white; dorsal and anal with rows of red blotches; gill-cover with a spot. Inhabits muddy waters of North Queensland down to the Capricorns. To 38 centimetres.

Family Carapidae: Pearl-fishes or Messmates

Hump-back Messmate *Carapus houlti* PLATE 125:1

Colour greyish-brown, dotted darker, the dots extending on to the pectoral-base. Known only from two examples trawled in 36 fathoms off Double Island Point in 1922 (28.3 and 23.6 centimetres). When caught they were enclosed in the remains of a holothurian or bêche-de-mer.

Family Nomeidae: Pomfrets

Portuguese Man-o'-war Fish *Nomeus albula* PLATE 125:2

Colour brownish or blackish above or sometimes blue, with large brown blotches and spots over body and fins. An open-sea fish found in temperate and tropical seas. Its name is derived from its habit of swimming beneath the Portuguese Man-o'-war (Physalia). Length 15 centimetres.

Whitelegge's Blubber-fish *Psenes whiteleggii* PLATE 125:3

Colour delicate salmon, with three brownish transverse bands, the first very broad; head brown above, silvery below, including gill-covers. A small fish, usually found in open-sea, but has been taken in Moreton Bay, southern Queensland. It grows to a length of about 5 centimetres. Uncommon.

Family Sphyraenidae: Barracudas or Sea-pikes

Pick-handle Barracuda *Sphyraena jello* PLATE 54:2

Colour brownish or greenish above, silvery-white below, the colour of the back connecting to a number of cross-bars of the same colour; lower lobe of tail yellowish; ventrals yellowish. A vicious fish found commonly in the waters of North Queensland. Grows to about 2 metres.

Striped Barracuda *Sphyraena obtusata* PLATE 54:3

Colour greyish-green or yellowish-green above, silvery-white below; the colour of the upper parts descends in a festooned edge below the lateral line; fins yellowish; tail bright yellow in the young. Common on the Queensland coast. Grows to at least 50 centimetres. Good food-fish.

Family Mugilidae: Mullets

Shark Mullet *Rhinomugil nasutus* PLATE 54:4

Colour silvery. Recognized from all other mullets by its pointed and flattened head and the small eyes on top of the head. Frequents shallow waters in muddy areas from Rockhampton northwards. Swims in small schools of twelve or so, with the head and eyes out of water. To 30 centimetres.

Tallegalane or Sand Mullet *Myxus elongatus* PLATE 125:4

Colour is reddish-brown or dark greenish above, the sides pink or silvery, white below; a black axillary spot at pectoral base; a golden spot on upper edge of gill-cover. A common southern fish frequently caught in southern Queensland. A good food-fish. Grows to 40 centimetres.

Freshwater Mullet *Trachystoma petardi* PLATE 125:5

Colour is greenish-black above and silvery below; fins black; eye bright orange-yellow. A large freshwater mullet which lives and breeds in fresh water. It is fairly common in the streams and lagoons of southern Queensland. Grows to nearly 60 centimetres. Flesh of fair quality.

Fantail Mullet *Mugil georgii* PLATE 126:1

Colour pale olive-brown, sides and below silver with golden tinge; a black spot in pectoral axil; pectorals, anal, and ventrals yellow. Identified by the thick upper lip and slightly emarginate tail. Coasts and estuaries of Queensland. Common and good food-fish. To 25 centimetres.

Long-finned Mullet *Liza strongylocephalus* PLATE 126:2

Colour is greenish above and silvery below; tail with a dark margin; pectoral base with a jet-black mark; upper part of bone between cheek and gill-cover bright golden. A small, wide-ranging and common mullet in the waters of North Queensland. Good food-fish. To about 25 centimetres.

Flat-tail Mullet *Liza dussumieri* PLATE 126:3

Colour of back greyish-green, brownish on head; sides silvery, washed bronze, white below; longitudinal bands of golden-brown on back and sides. A common Queensland mullet and a fine food-fish. Said to be the best of all mullets, with flesh of delicate flavour. To 38 centimetres.

Tade Mullet *Liza tade* PLATE 127:1

Colour pale olive, whitish below; body with five to seven indistinct longitudinal lines along the scale-rows; no dark spot at pectoral base. A common mullet which frequents the coastal waters of North Queensland. It attains a length of at least 45 centimetres and is a very good food-fish.

Sea Mullet *Mugil cephalus* PLATE 127:2

Steel-blue, tinged greenish or olive, silvery below. Young spend first few years of their lives far up freshwater streams, moving down to sea at three years old for the ripening of the ova. In the months of May and June they are netted in great numbers in southern Queensland. To 76 centimetres.

Diamond-scaled Mullet *Liza vaigiensis* PLATE 55:1

Colour dark greenish or blackish above, lighter on the sides and silvery below; scales with blackish borders and longitudinal blackish bands on the sides. A heavy-built mullet common on the Queensland coast. Grows to at least 60 centimetres and about 4.5 kilograms. Good food-fish.

Blue-tail Mullet *Valamugil seheli* PLATE 55:2

Colour details as shown in plate. Readily identified by the soft dorsal fin and bright light blue tail. The largest of the Queensland mullets. Inhabits North Queensland in vast shoals, where it grows to over 10 kilograms. South of Rockhampton it is only met with as a straggler.

Warty-nosed Mullet *Crenimugil crenilabis* PLATE 127:3

Colour of head, body, and tail greyish-green, silvery on sides and below; fins greyish, except pectorals, which are yellowish with axil spot of blue-black. A common fish in North Queensland where it swims in schools, as in most species. Grows to 25 centimetres. A good food-fish.

Ramsay's Mullet *Liza ramsayi* PLATE 127:4

Colour silvery all over, with a bluish tinge on the upper surface. A little known mullet from the Burdekin River district, North Queensland. It may be recognized by its thick upper lip, eleven anal rays, and the absence of adipose eyelids. It grows to 25 centimetres.

Basket Mullet *Liza diadema* PLATE 128:1

Colour is greyish above and silvery on the sides and below, each scale with a dark edge and a horizontal whitish streak. A large mullet which inhabits northern Australia from the Gulf of Carpentaria to north-west Australia. It grows to 76 centimetres and is a good food-fish.

Tiger Mullet *Liza argentea* PLATE 128:2

Colour steel-blue above, passing through silver to white below; scales with narrow longitudinal streaks forming bands; a small black axillary spot present, preceded by a golden blotch. A very valuable food-fish in southern Queensland. Grows to about 45 centimetres. Common.

Family Atherinidae: Silversides and Hardyheads

Marine Blue-eye *Pseudomugil signifer* **PLATE 128:3**

Colour olivaceous, each scale with a darker edge; sides and belly white; eye blue, the pupil black; second dorsal largely orange-yellow and attenuated in males. Inhabits shallow water near mangrove-roots in North Queensland in thousands. Male very pretty. Grows to 5 centimetres.

Common Hardyhead *Pranesus ogilbyi* **PLATE 128:4**

Colour pale green, each scale edged and dotted with black; white below, with a series of black dots below the lateral line; a silvery lateral band along the side to tail-base. A common little fish which swims in great shoals in shallow Queensland bays. Grows to 17 centimetres.

Family Polynemidae: Threadfins or Tassel-fishes

Giant Threadfin *Eleutheronema tetradactylum* **PLATE 55:3**

Colour details as shown in plate. A large and valuable food-fish, common on the coast and in the bays and estuaries of Queensland. Also ranges outside Australia. In India it is said to reach 1.75 metres and 145 kilograms, but average Queensland fish are about 58 centimetres and 2 kilograms.

Sheridan's Threadfin *Polynemus sheridani* **PLATE 56:1**

Colour details as shown in plate. Frequents the whole coastline of Queensland north of Maryborough. A valuable food-fish with flesh of excellent flavour. Fish of 1 metre and over, weighing 18 kilograms gutted, are common at Maryborough. It is said to reach a weight of 45 kilograms.

Striped Threadfin *Polynemus plebeius* **PLATE 56:2**

Colour golden, with a greyish tinge above; dark brown longitudinal lines between the scale-rows; pectorals dull yellow. A wide-ranging fish known in Australian waters only from Queensland and New South Wales, but is somewhat rare in both States. Grows to about 60 centimetres.

Flat Salmon *Polynemus multiradiatus* PLATE 56:3

Colour golden on head, back, and post abdomen; silvery-pink on belly, the two colours sharply defined; fins yellowish, the pectorals densely speckled with black. A very common fish found in the rivers and on the coast of Queensland. Good food-fish. Grows to 28 centimetres.

Family Scorpaenidae

Red Scorpion Cod *Scorpaena cardinalis* PLATE 129:1

Colour carmine-red, pearly-white below; under head is pale red, marbled yellow; spinous dorsal red with silvery blotches. One of the most important of the scorpion cods, has high value as a food-fish. Inhabits southern Queensland, over rocky shoals. Length to 45 centimetres.

Little Scorpion Cod *Scorpaena bandanensis* PLATE 129:2

Colour is greenish above and rosy-green to golden below, clouded or variegated with brown; fins variegated with rosy-brown and with dark bands. Although widely distributed outside Australia, this little fish is apparently uncommon in Queensland. It grows to about 10 centimetres.

Marbled Coral Cod *Scorpaena bynoensis* PLATE 57:1

Colour variable, from between greyish and blackish to bright brown and white; roseate below and on base of tail; fins pale, barred and mottled reddish or brown. A common little fish in the coral reefs of Queensland. Can inflict painful stings with its spines. To 14 centimetres.

Guam Scorpion Cod *Scorpaenodes guamensis* PLATE 129:3

Colour rich reddish or dusky-brown with darker bars and spots on the head, body, and fins; a more or less distinct black spot on the gill-cover. A small stinging fish found under the slabs of dead coral clinker on the Great Barrier Reef of Queensland. It grows to 10 centimetres.

Bullrout *Notesthes robusta* PLATE 57:2

Colour details as shown in plate. Although originally a marine form, this fish has taken to a brackish and even freshwater existence. It is commonly found in rivers that discharge into the sea in southern Queensland. Causes painful wounds with its spines. Edible. To 28 centimetres.

Marbled Fortesque *Centropogon marmoratus* PLATE 129:4

Colour yellowish, marbled with brown. A common little fish in Moreton Bay, southern Queensland, where it frequents weedy flats in shallow bays. It can inflict severe stings with its very sharp spines, particularly those just before the eye. It grows to 11 centimetres.

Red Fire-fish *Pterois volitans* PLATE 58

Colour details as shown in plate. A very handsome reef-dwelling fish of great beauty when seen among the corals on the Great Barrier Reef. The dorsal spines are long and needle-like and charged with venom, the fish thus being able to inflict painful wounds. Grows to 38 centimetres.

Zebra Fire-fish *Dendrochirus zebra* PLATE 129:5

Colour is reddish, with dark reddish-black cross-bands; eye with radiating streaks; soft dorsal, anal, and caudal with dark spots; a yellowish-white spot in the pectoral axil. A beautiful but venomous fish met with in the coral reefs of North Queensland. Grows to 18 centimetres.

Ocellated Wasp-fish *Hypodytes carinatus* PLATE 129:6

Colour lilac above, lined and marbled darker, shading through lavender and silver to white belly; a large black, white-edged spot on ninth to eleventh dorsal spines and two much smaller ones at first to third spines. Not usually met with except when trawled. Grows to 16 centimetres.

Short-spined Wasp-fish *Apistops caloundra* PLATE 129:7

A broad, white-edged black blotch between the ninth and twelth spines of the dorsal fin. Until recently this little fish was known only from the type specimen. It has now been taken fairly commonly by the many prawn-trawlers at several localities in Queensland. Grows to 16 centimetres.

Black-banded Wasp-fish *Minous versicolor* PLATE 129:8

Colour of body reddish-pink, strongest on breast and belly; upper half of body with irregularly-shaped markings of blackish-brown which extend on to the dorsal fins. Not commonly met with unless taken by trawl; they are numerous down to 35 fathoms. Grows to 11 centimetres.

Brown Carpet-fish *Peristrominous dolosus* PLATE 129:9

Colour of head and body light brown, with darker marblings, forming four obscure bars on the back; fins variegated with grey and brown; the pectorals and anal very dark. A small fish known only from southern Queensland where it has been taken by trawl. It grows to 8 centimetres.

Stone-fish *Synanceia horrida* PLATE 130:1

Colour variable; usually brownish, with greyish blotches. This deadly fish is common on the shallow mud-flats of southern Queensland as well as on the coral reefs of North Queensland. It can inflict dangerous wounds with its many venom-laden spines. Grows to a length of 60 centimetres.

Monkey-fish *Erosa erosa* PLATE 130:2

Colour brown, with a rosy tinge on shoulders; on the side, above and beneath the adpressed pectoral, is a large patch, roseate above and yellow below. A little-known fish recorded from the Queensland coast. These fish were trawled down to 33 fathoms. Grows to 13 centimetres.

Bearded Ghoul *Inimicus barbatus* PLATE 130:3

Colour is black or brown above, passing into dark brown or light brown on the sides and usually freckled with blackish-brown; the belly is yellowish-brown. A grotesque fish commonly taken along the coast of Queensland, mainly by the prawn-trawlers. Grows to at least 33 centimetres.

Rogue-fish *Tetraroge leucogaster* PLATE 130:4

Colour is brownish, marbled with brown, the spots extending on to the dorsal fin; the abdomen is white. A small and stinging fish which is very rarely met with and then only in the waters of North Queensland and in Western Australia. Should be handled with care. Length 6 centimetres.

Freckled Wasp-fish *Paracentropogon vespa* PLATE 130:5

Colour chestnut above, freckled with lavender; lavender below, blotched with chestnut; wavy chestnut bars on fins and tail; ventrals blackish, membrane of inner ray white. Has been taken in Queensland by trawl at depths from 7-13 fathoms. Rarely met with. Grows to 8 centimetres.

Yellow Wasp-fish *Liocranium scorpio* PLATE 130:6

Colour brown, sides lilac; scattered black light-edged spots above lateral line; belly white, rest of body powdered with black spots; the spinous dorsal lavender, spines ringed with black. Little-known; all recorded examples taken by trawl in North Queensland. Length 10 centimetres.

Family Aploactinidae

Port Curtis Mossback *Bathyaploactis curtisensis* PLATE 130:7

Colour is brownish, with darker markings and most stongly marked on the fins; around the eye are dark brown bars with well-marked margins. It is commonly taken by the prawn-trawlers in southern Queensland. Also found in northern waters. Grows to about 6 centimetres.

Northern Mossback *Paraploactis trachyderma* PLATE 130:8

Colour dark brown, with orange-yellow markings; snout and lips are orange; dorsal has orange patch over first three spines and membranes and on last five rays and membranes. A small fish commonly taken by trawlers in Moreton and Deception bays, southern Queensland. Length 15 centimetres.

Sandpaper-fish *Adventor elongatus* PLATE 130:9

Colour is brownish-grey all over, somewhat lighter below; all fins are somewhat darker than the body, being blackish-grey, the dorsals and tail are mottled. This fish is known only from a few examples taken by trawls on the coast of Queensland. The largest was about 10 centimetres.

Family Pataecidae

Red Indian Fish *Pataecus fronto* PLATE 130:10

Colour is red, brick-red, or orange-red, with or without four black spots on the upper part of the body. This remarkable fish, which bears a curious resemblance to a Red Indian head-dress, is only occasionally met with in southern Queensland. It grows to about 23 centimetres in length.

Family Dactylopteridae: Flying Gurnards

Purple Flying Gurnard *Dactyloptena orientalis* PLATE 59

Colour details as shown in plate. A most handsome fish commonly found along the whole of the coastline of Queensland. Prawn-trawlers frequently take them in their catches off the rivers and in the bays and ocean. These fish glide short distances only. Grow to about 40 centimetres.

Rufous Flying Gurnard *Dactyloptena papilio* PLATE 60:3

Colour details as shown in plate. Known only from Queensland, and although originally described by Ogilby in 1910, the species was not encountered again until the years 1952-55, when the prawn-trawlers began to collect them. They grow to a length of at least 18 centimetres.

Family Triglidae: Gurnards

Long-finned Gurnard *Lepidotrigla calodactylus* PLATE 60:1,2

Colour details as shown in plate. A little-known fish seldom taken and then only by trawlers operating along the Queensland coast, where it has been caught at depths down to 70 fathoms. Further trawling will, no doubt, reveal it to be common in deeper water. Grows to 18 centimetres.

Butterfly Gurnard *Paratrigla papilio* PLATE 131:10

Colour is brownish, whitish below; head yellowish; spinous dorsal yellow and with a large black spot edged with white. A common fish, but seldom seen except when taken in the nets of the trawlers along the Queensland coast at depths of 12-26 fathoms. Grows to 18 centimetres.

Red Gurnard *Chelidonichthys kumu* PLATE 131:1

Colour red, reddish, grey, or olive; sides reddish-yellow, washed with pink; cheeks scarlet; dorsals red and yellow; pectorals with a large black blotch and blue spots. Uncommon in Queensland and seldom captured, except by outside fishermen. Good food-fish. To 53 centimetres.

Family Hoplichthyidae: Spiny Flatheads

Ogilby's Spiny Flathead *Hoplichthys ogilbyi* PLATE 131:2,3

Colour pink or yellow, with indefinite darker spots and lines; back crossed with indistinct bands; first dorsal with wavy blackish cross-lines. Known only from two examples trawled by the *Endeavour* off Cape Moreton, southern Queensland, in 1910. Size 1.5 and 1.97 centimetres. Not seen since.

Family Platycephalidae: Flatheads

Fringe-eyed Flathead *Platycephalus nematophthalmus* **PLATE 61:1**

Colour details as shown in plate. A very common fish in Queensland waters where it is taken in the nets of the prawn-trawlers. Otherwise met with only occasionally. A good food-fish which grows to about 35 centimetres. It can be identified by the large, deep pit behind each eye.

Dwarf Flathead *Elates ransonetti* **PLATE 61:2**

Colour details as shown in plate. This small, slender fish is very occasionally taken at Townsville, North Queensland, by the trawlers, prior to which it had not been seen since 1910, when four were trawled at Pine Peak and Cape Gloucester, Queensland. It grows to about 20 centimetres.

Dusky Flathead *Platycephalus fuscus* **PLATE 131:4**

Colour is dark brownish, sometimes with a few lighter spots; sides of head with numerous dark spots; fins and tail spotted, the latter with a large dark area. Lives in the estuaries and on shallow flats along the Queensland coast. A valuable food-fish. Length to 1.25 metres.

Sand Flathead *Platycephalus arenarius* **PLATE 131:5**

Colour light yellowish-brown or sandy, closely speckled with fine spots, spines and rays with chestnut bands, the membranes hyaline; tail white with black stripes. A lover of the sandy flats and bays of Queensland. Good food-fish. Grows to 45 centimetres; average is 30 to 35 centimetres.

Harris's Flathead *Platycephalus harrisii* **PLATE 131:6,7**

Colour light brown, passing into greyish-white below; dorsal spines and rays with brown spots; anal whitish; tail with large blackish-brown spots largest on the lower half of the fin. A small fish commonly taken by the prawn trawlers on the Queensland coast. Grows to 23 centimetres.

Rock Flathead *Platycephalus cirronasus* PLATE 131:9

Broad blackish-brown cross-bands on the back, much wider than the narrow fawn-coloured interspaces; these interspaces break on the lower sides into yellowish wavy lines and paler brown spots. Inhabits the weed-covered rocks in southern Queensland. It grows to about 38 centimetres.

Family Pegasidae

Short-tailed Dragon-fish *Pegasus volitans* PLATE 131:8

Colour is brownish, finely reticulated with darker brown; snout and the last two or three tail-rings black; pectorals with broad milky-white edges and sometimes a white band. Found along the coast of Queensland but it is not commonly met with. It grows to only about 10 centimetres long.

Long-tailed Sea Dragon *Parapegasus natans* PLATE 61:3

Colour details as shown in plate. This is a very common and curious little fish which is found along the whole coastline of Queensland. Usually collected along the many beaches among after-storm litter. It grows to 18 centimetres and is readily identified by the long flattened tail.

Family Psettodidae

Queensland Halibut or Adalah *Psettodes erumei* PLATE 62:1

Colour details as shown in plate. Juveniles are blackish, with narrow white transverse bands. The largest flat-fish in Australian seas. Common on the North Queensland coastline, but seldom taken by line. Grows to 60 centimetres and a weight of 9 kilograms. Excellent food-fish.

Family Bothidae: Flounders

Large-toothed Flounder *Pseudorhombus arsius* PLATE 62:2

Colour details as shown in plate. This is one of the most common of the

Australian flat-fishes. It is a valuable food-fish with flesh of excellent flavour. Found along the whole coastline of Queensland and is wide-ranging outside Australia. It grows to at least 38 centimetres.

Small-toothed Flounder *Pseudorhombus jenynsii* PLATE 132:1

Colour brownish or greyish, with darker spots and markings and with five eye-like spots on body. A common fish along the Queensland coast where it has been taken as far north as Townsville. It grows to a length of 40 centimetres and is a fine food-fish with flesh of good flavour.

Fison's Flounder *Arnoglossus fisoni* PLATE 132:2

Colour sandy-grey, with numerous small, faint, eye-like spots, a narrow yellowish band, parallel with dorsal and anal profiles along the base of the interspinous rays; fins with dark and light spots. It inhabits the coast of southern Queensland. Taken by trawlers. Length 13 centimetres.

Crested Flounder *Lophonectes gallus* PLATE 132:3

Colour is brownish or greyish, with or without darker markings, the most conspicuous of which are three blotches on the lateral line; fins dotted. A southern fish which straggles into southern Queensland. It grows to only 20 centimetres. The anterior dorsal rays are prolonged in the males.

Spotted-tail Flounder *Engyprosopon grandisquama* PLATE 132:4

Colour is light brownish, mottled with darker markings; fins with small dark brown spots; a pair of large blackish spots on the tail. It has been taken by prawn-trawlers from various parts of the coast of Queensland at depths from 5 to 30 fathoms. It grows to about 14 centimetres.

Panther Flounder *Bothus pantherinus* PLATE 132:5

Colour is brownish, with paler and darker spots, blotches, rings, or eye-like spots; usually a large dark blotch on the middle of the straight portion of the lateral line. Inhabits Queensland in the far northern waters where it is quite common. A good food-fish. Length to 45 centimetres.

Feather-fin Flounder *Grammatobothus pennatus* PLATE 132:6

Colour is brownish, with three prominent eye-like spots as a triangle, the one on the straight portion of the lateral line being darker. It is known only from eastern Queensland where it has been trawled in depths from 19 to 33 fathoms. Length 20 centimetres. Identified by its dorsal rays.

Family Pleuronectidae

Cockatoo Flounder *Samaris cacatuae* PLATE 132:7

Colour lavender, dotted with black and with three series of rather obscure black spots, more or less encircling a central lavender spot. Anterior rays of dorsal fin greatly prolonged. A rare fish trawled in 1910 by the *Endeavour* off Gloucester Head, Queensland. Length 17 centimetres.

Family Soleidae: True Soles

Tiger Sole *Aesopia heterorhinos* PLATE 133:1

Colour rich brownish-olive, with irregular vertical bands, blotches, and spots, edged with black, continued on to the vertical fins, which are often edged with black or with a submarginal band of black. It has been recorded from Torres Strait, North Queensland. Grows to 14 centimetres.

Peacock Sole *Pardachirus pavoninus* PLATE 63:1

Colour details as shown in plate. A very beautiful and showy species which inhabits the waters of North Queensland where it is fairly commonly met with. It is readily recognized by its distinctive colour pattern. It grows to at least 21 centimetres and is a very good food-fish.

Narrow-banded Sole *Aseraggodes macleayanus* PLATE 133:3

Colour lavender-grey, with narrow, irregular, brown cross-bands. Inhabits the coastal waters of southern Queensland, particularly Moreton Bay. Enters our inlets and bays in goodly numbers during spring months. Has flesh of excellent flavour. Grows to about 28 centimetres.

Black Sole *Achylopa nigra* PLATE 133:2

Colour blackish or olive-brown, sometimes with small yellowish or creamy-white spots; fins stained with orange or straw-colour. Prefers the muddy estuaries and bays along the coast of Queensland rather than the more sandy areas. A splendid food-fish. Common. Grows to 35 centimetres.

Sharp-headed Sole *Phyllichthys sclerolepis* PLATES 63:2, 133:5

Colour details as shown in plate. The ventrals are more or less united and the right ventral is joined to the first anal ray. A sole which inhabits the waters of the Northern Territory and Queensland as far south as Moreton Bay. It grows to 28 centimetres and is a good food-fish.

Family Cynoglossidae: Tongue Soles

Lemon Tongue Sole *Paraplagusia unicolor* PLATE 133:4

Colour pale yellowish-brown, with round spots of an even lighter shade; fins are profusely ornamented with brown and milk-white spots; bluish-white below. Common on sandy bottoms in shallow water in southern Queensland. Grows to 33 centimetres. Good food-fish. Has two lateral lines.

Two-lined Tongue Sole *Cynoglossus bilineatus* PLATE 64:1

Colour details as shown in plate. Inhabits the waters of the east coast of North Queensland where it is very common along the mainland beaches where it is taken in seine nets. Has two lateral lines on both sides of the body. A very good food-fish, growing to about 40 centimetres.

Family Echeneidae: Sucking-fishes

Slender Sucking-fish *Echeneis naucrates* PLATE 64:2

Colour details as shown in plate. A very common fish which seems to prefer, almost exclusively, to "hitch-hike" along on sharks, and to which it attaches itself by the powerful sucking-disk situated on top of its head. Grows to at least 1 metre and found in all warm seas.

Short Sucking-fish *Remora remora* PLATE 133:6

Colour is blackish or brownish, the colour being nearly uniform above and below. A short, thick-set sucking-fish not so commonly seen as the previous species. Found in all warm and temperate seas. Usually found attached to large sharks and turtles. Grows to at least 45 centimetres.

Family Triacanthidae: Triple-spines or Horn-fishes

Black-finned Triple-spine *Triacanthus biaculeatus* PLATE 64:3
Colour details as shown in plate. A common fish inhabiting the whole Queensland coastline. It is commonly captured by seine net on sandy beaches. No value as a food-fish; the flesh is unpalatable and eating it could be harmful. Grows to about 24 centimetres in length.

Family Balistidae: Trigger-fishes

Big-spotted Trigger-fish *Balistoides conspicillum* PLATE 65:1
Colour details as shown in plate. A strikingly handsome species which frequents rocky situations along the coast of Queensland but is not commonly met with by casual observers. It grows to 33 centimetres. The flesh is poisonous and can cause violent illness and perhaps death.

Yellow-spotted Trigger-fish *Pseudobalistes fuscus* PLATE 134:1
General colour bright bluish-purple, each scale with a yellow spot; snout, chin, and throat violet; upper lip with two yellow lines; yellow spots on dorsal fins. This Indo-Pacific species is rare in Queensland. It has been taken in southern Queensland. Grows to 50 centimetres. Poisonous.

Flat-tailed Trigger-fish *Abalistes stellaris* PLATE 65:2
Colour details as shown in plate. A common fish, very variable in colour. It is found along the whole Queensland coastline. Differs from all other trigger-fishes by the peculiar shape of the stalk of the tail, which is flattened above and below. Length 50 centimetres. Poisonous flesh.

Red-lined Trigger-fish *Balistapus undulatus* PLATE 66:1
Colour details as shown in plate. A beautiful fish commonly found on the coral reefs of North Queensland. It grows to a length of at least 35 centimetres. All such trigger-fishes should be avoided as food, for the flesh is poisonous and can cause violent illness and perhaps death.

Spotted Trigger-fish *Canthidermis rotundatus* PLATE 66:2
Colour details as shown in plate. A wide-ranging species which has been

recorded from Torres Strait in far North Queensland and also in the Brisbane River, southern Queensland, both examples being juveniles. Grows to 40 centimetres. It should not be eaten; the flesh is poisonous.

Family Monacanthidae: Tape-fishes, File-fishes and Leather-jackets

Beardie or Tape-fish *Anacanthus barbatus* **PLATE 67:1**

Colour details as shown in plate. A curious fish found along the whole of the coastline of Queensland, both on the coral reefs and along the mangrove-lined shores and mud-banks. When laid upon a beach it can throw itself one foot into the air. Length 31 centimetres. Poisonous.

Prickly Leather-jacket *Chaetoderma penicilligera* **PLATE 67:2**

Colour details as shown in plate. A fairly common fish on the coast of Queensland from north to south, though it is more frequently met with in the north. It is readily recognized by its colour-pattern and by the large threads of skin. Grows to 25 centimetres. Flesh poisonous.

Fan-bellied Leather-jacket *Monacanthus chinensis* **PLATE 68:1**

Colour details as shown in plate. An extremely common little fish which is met with all along the coast of Queensland. It grows to about 28 centimetres. Like most members of the group its flesh is poisonous and must not be eaten. Flesh of others of the family has caused death.

Chinaman Leather-jacket *Nelusetta ayraudi* **PLATE 134:2**

Colour yellowish-brown; head tinged blue; sides and abdomen pale yellowish-brown; all fins bright yellow. A large fish which is only a straggler into Queensland from New South Wales. Curiously enough this particular species is a good food-fish. It grows to 50 centimetres.

Beaked Leather-jacket *Oxymonacanthus longirostris* **PLATE 68:2**

Colour details as shown in plate. This brightly-coloured little leather-jacket can often be observed swimming amongst the living corals along the edges of the reefs adjacent to the deeper waters of the Great Barrier Reef. It grows to a length of at least 8 centimetres.

Unicorn Leather-jacket *Alutera monoceros* PLATE 134:3

Colour is brownish or blackish, the fins yellow. A rare fish on the coast of Queensland, its inclusion in the State's fauna being based on a specimen which was taken off Caloundra, southern Queensland. It inhabits all warm seas and is said to reach a length of 76 centimetres.

Figured Leather-jacket *Alutera scripta* PLATE 69:1

Colour details as shown in plate. This large and handsome species so far (1966) has been recorded in Queensland only from the waters north of Townsville. It occurs in all warm seas and is a dangerous fish to eat because of its poisonous flesh. Grows to about 1 metre.

Family Ostraciontidae: Box-fishes, Cow-fishes, Turret-fishes

Robust Box-fish *Strophiurichthys robustus* PLATE 134:5

Colour yellowish-brown, with large dark spots on the back and the upper sides. This fish inhabits the waters of Australia and the tropical Pacific at depths down to 150 fathoms. Fairly commonly taken by the prawn-trawlers at Coolangatta, southern Queensland. Length 25 centimetres.

Long-nosed Box-fish *Rhynchostracion nasus* PLATE 69:2

Colour details as shown in plate. Young are bright orange-yellow, with numerous black spots. This fish is apparently rare on the coast of Queensland. It ranges through the Pacific to the Malay Peninsula and the Indian Ocean. Found also in Western Australia. Grows to 21 centimetres.

Blue-spotted Box-fish *Ostracion cubicus* PLATE 69:3

Colour details as shown in plate. Extremely common in Queensland reef-waters from the far north to the south as far as and including the Capricorn and Bunker groups. It is a very fast swimmer and is said to attain a length of 45 centimetres. A wide-ranging fish outside Australia.

Cow-fish *Lactoria cornutus* PLATE 70:1

Colour details as shown in plate. Commonly seen everywhere on the Great Barrier Reef as well as other localities along the coastline. When the fish cruises slowly, the great tail is folded, if it speeds away frightened, the tail is fully expanded. Grows to 50 centimetres.

Turret-fish *Rhinesomus reipublicae* PLATE 134:4

Scattered spots, dark or milky-white, on head, body, and tail-base; usually some dark blotches on the back and sides; sometimes dark stripes on the cheeks. An open-sea species seen only when it is cast up on southern Queensland beaches or when taken by the trawlers. To 23 centimetres.

Spiny-backed Cow-fish *Lactoria fornasini* PLATE 134:7

Body with some irregular bluish and blackish markings. A tropical species recorded from Australia on a single example which was taken many years ago at Southport, southern Queensland. It has not been seen here since. Identified by the spine before the eye. Grows to about 15 centimetres.

Family Tetraodontidae: Toadoes

Weeping Toado *Torquigener pleurogramma* PLATE 134:6

Colour greyish or greenish above, with dark brown reticulations that form spots; four broad dark bands across the back; white below; black bars down the cheeks. Not common in Queensland but occasionally taken by trawlers in southern Queensland. Flesh poisonous. Length 18 centimetres.

Banded Toado *Torquigener pleurostictus* PLATE 70:2

Colour details as shown in plate. Common in the estuaries of the Northern Territory and Queensland. Although some species are used as food by native peoples, after removal of skin and entrails, there are many deaths through eating toadoes. Grows to a length of 18 centimetres.

Common Toado *Torquigener hamiltoni* PLATE 134:9

Colour brownish, with numerous close-set round black spots; cheeks with some brown vertical bands and spots; some large dark blotches on the sides; an indistinct dark band usually present across back; white below. Very common in Queensland. Length 14 centimetres. Flesh is poisonous.

Stars and Stripes Toado *Arothron hispidus* PLATE 70:3

Colour details as shown in plate. Common all along the Queensland coast. Hated by line-fishermen because of their habit of biting through hooks with their beak-like teeth. Grows to at least 50 centimetres. Poisonous. All toadoes can inflate their bodies with air and water.

A Toado *Arothron stellatus* PLATE 71:1

Colour details as shown in plate. Very variable. An Indo-Pacific fish found all along the Queensland coast. Grows to 60 centimetres. The poisonous flesh of all toadoes is said to be due to a dangerous alkaloid most abundant and virulent in the roe, milt, and gall-bladder.

Marbled Toado *Chelonodon patoca* PLATE 71:2

Colour details as shown in plate. In Australia this fish is found in Queensland and the Northern Territory. Also found in the tropical parts of the Indo-Pacific. It grows to a length of at least 38 centimetres. The teeth of toadoes are fused to form a parrot-like "beak". Poisonous.

Striped Toby *Canthigaster coronata* PLATE 134:8

Back with three to four blackish-brown cross-bands; around the eye and on the cheeks are zebra-like stripes; faint lines across the snout; breast with pale transverse bands. This very pretty fish is commonly met with only on the coral reefs of Queensland. Grows to 13 centimetres.

Family Diodontidae: Porcupine-fishes

Porcupine-fish *Tragulichthys jaculiferus* PLATE 71:3

Colour details as shown in plate. Found along the whole Queensland coastline. An extremely common fish in Moreton Bay during the cooler months, when dozens are taken in the seine nets of the fishermen. The flesh is highly poisonous. Grows to at least 25 centimetres. Body inflatable.

Long-spined Porcupine-fish *Diodon holocanthus* PLATE 135:1

Colour brownish above, dirty whitish below, with broad brown bands and spots; eyes greenish; fins hyaline or yellowish. Body inflatable, but not to the same extent as in the toadoes. Widely distributed in all warm seas. Rarely met with in Queensland waters. Grows to 50 centimetres.

Myers' Porcupine-fish *Dicotylichthys myersi* PLATE 135:2

Colour uniform olive-brown; lavender or white below; lower sides and belly with small black spots; a black vertical band below eye, a second in front of gill-opening and a third below the tip of the adpressed pectoral. Bays and inlets of southern Queensland. Grows to 43 centimetres.

Family Molidae: Sunfishes or Head-fishes

Ocean Sunfish *Mola mola* PLATE 135:4

Colour is dark grey above and greyish-brown on the sides, with reflections of silver; a light grey band runs from the base of the dorsal fin around the tail-base to the anal base. Length to 3 metres, and a weight of over 3.5 tonnes. Fairly common off the Queensland coast Harmless.

Oblong Sunfish *Ranzania laevis* PLATE 135:3

Colour dark blackish-brown; sides and belly silvery, with pattern as shown in plate. A small sunfish which grows only to about 76 centimetres. Although wide-ranging and found in Australian waters (Victoria and Western Australia) it has not yet been recorded from Queensland.

Family Gobiesocidae: Cling-fishes

Bridled Cling-fish *Lepadichthys frenatus* PLATE 135:5

Colour (when preserved) is pinkish-yellow; edge of upper lip scarlet, continued as a streak backward through eye towards margin of gill-cover; fins colourless. A South Pacific fish rarely seen in Queensland but it has been taken in the Capricorn and Bunker groups. Grows to 5 centimetres.

Family Batrachoididae: Frogfishes

Frogfish *Batrachomoeus dubius* PLATE 135:8

Colour dark brown, blotched and marbled lighter; pale brown below, blotched with bluish-white; fins brown, with spots and bars of lighter colour. Common on the east coast of Queensland, in river-mouths and muddy foreshores, also in deep off-shore waters. Length 35 centimetres.

Banded Frogfish *Halophryne diemensis* PLATE 135:6,7

Colour varies from violet to purplish-black; pearl-grey or lilac below; cross-bands and mottlings of lighter colour, sprinkled with spots and dots and sometimes with scarlet or orange above. Very variable in colour. Common on the Queensland coast. Grows to 26 centimetres.

Family Lophiidae: Fishing-frogs

Broad-headed Fishing-frog *Lophiomus setigerus* PLATE 136:1,2

Colour pale lilaceous-brown above, the skin filaments and the tentacles darker; white below. An apparently rare fish known only from a few specimens taken by trawl off the coasts of southern Queensland and New South Wales at depths from 40-73 fathoms. Length about 19 centimetres.

Family Ogcocephalidae: Sea-bats

Spiny Sea-bat *Halieutaea brevicauda* PLATE 136:3

Colour lavender above, tail and edges of disk shading to grey. A grotesque angler first known from an example trawled in 1910 by the *Endeavour* off Cape Moreton, southern Queensland in 73 fathoms. Others since trawled in New South Wales and Tasmania. Rare. Grows to 15 centimetres.

Family Antennariidae: Angler-fishes

Marbled Angler *Histrio histrio* PLATE 72

Colour details as shown in plate. Skin of the body and dorsal fins with fleshy appendages or tags, which can be absent in the young. A wide-ranging open-sea fish which lives in masses of floating seaweeds and is sometimes cast up on our beaches. Grows to a length of 13 centimetres.

Smooth Angler *Histiophryne bougainvilli* PLATE 136:5

Colour yellow, with faint greyish reticulations and a few large and irregular rusty patches on the sides; eyes green. An Australian species known in Queensland for the first time from an example caught in the Capricorns at Heron Island. It is rare here. Grows to 5 centimetres.

Striped Angler *Antennarius tridens* PLATE 136:4

Colour brownish or yellowish, with blackish streaks which radiate from the eye and break up into spots on the belly; the fins are also barred and spotted. A common angler found along the coast of southern Queensland but less frequently met with northwards. Grows to 30 centimetres.

Black Angler *Antennarius moluccensis* **PLATE 136:7**

Colour variable; usually dark brown or blackish, almost uniform or with some white or rose-coloured spots; pectoral and ventral fins are white-edged. Fairly common on the coast of Queensland. It grows to 30 centimetres. As in all anglers the stomach is capable of great distension.

Harlequin Angler *Tathicarpus muscosus* **PLATE 136:6**

Colour brilliant orange or brownish, densely spotted and streaked with blackish or greenish spots, bars, or eye-like spots. Colour variable. The first dorsal spine of anglers is fine and rod-like and ends in a "flesh-bait" which lures prey. Fairly common in Queensland. To 11 centimetres.

Index

Figures in bold type indicate plate numbers

COMMON NAMES

SCIENTIFIC NAMES

PLATE 1

1 

2 

Blue-spotted Stingaree, *Amphotistius kuhlii.*
1. Width 23cm;
2. Width 18cm.

PLATE 2

1

2

1. Coach-whip Ray, *Himantura uarnak*, width 30cm.
2. Blue-spotted Lagoon Ray, *Taeniura lymma*, width 18cm.

PLATE 3

1

2

3

4

1. Ox-eye Tarpon, *Megalops cyprinoides*, 45cm.
2. Bone-fish, *Albula vulpes*, 35cm.
3. Wolf Herring, *Chirocentrus dorab*, 35cm.
4. Whiskered Anchovy, *Thryssa setirostris*, 13cm.

PLATE 4

1

2

3

1. Hamilton's Anchovy, *Thryssa hamiltoni,* 6cm.
2. De Vis' Anchovy, *Stolephorus devisi,* 8.5cm.
3. Hairback Herring, *Nematalosa come,* 6.5cm.

PLATE 5

1 

2 

3 

4 

1. Hasselt's Sprat, *Dussumieria hasseltii*, 18cm.
2. Spotted Herring, *Harengula koningsbergeri*, 16cm.
3. Australian Pilchard, *Sardinops neopilchardus*, 20cm.
4. Black-spot Long Tom, *Strongylura strongylura*, 43cm.

PLATE 6

1
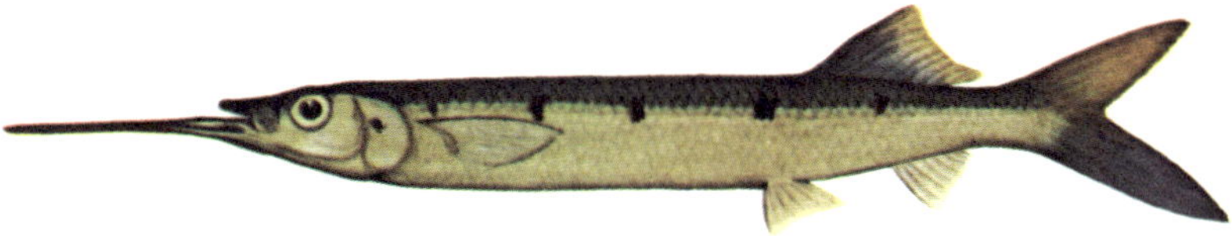

2

3

4

5

1. Black-barred Garfish, *Hemirhamphus far,* 40cm.
2. Three-by-two Garfish, *Hemirhamphus robustus,* 28cm.
3. Spoon-fin Garfish, *Zenarchopterus dispar,* 20cm.
4. Snub-nosed Garfish, *Arrhamphus sclerolepis,* 20cm.
5. Flute-mouth, *Fistularia commersonii,* 18cm.

PLATE 7

1

2

1. Scarlet-fin Soldier-fish, *Adioryx spinifer,* 35cm.
2. Crowned Soldier-fish, *Adioryx diadema,* 15cm.

PLATE 8

1 

2 

1. Red Soldier-fish, *Adioryx ruber*, 23cm.
2. Rock Flag-tail, *Kuhlia rupestris*, 25cm.

PLATE 9

1

2

1. Black-finned Cardinal-fish, *Apogon nigripinnis*, 11cm.
2. Cardinal-fish, *Apogon cooki*, 6cm.

PLATE 10

1

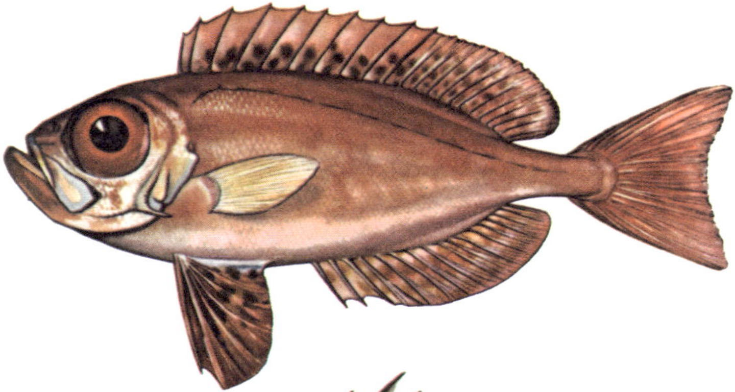

2

1. Red Bulls-eye, *Priacanthus macracanthus,* 15cm.
2. Rifle-fish or Archer-fish, *Toxotes chatareus,* 18cm.

1 

2 

1. Yellow Emperor, *Diploprion bifasciatum*, 20cm.
2. Lunar-tailed Rock Cod, *Variola louti*, 35cm.

PLATE 12

1

2

3

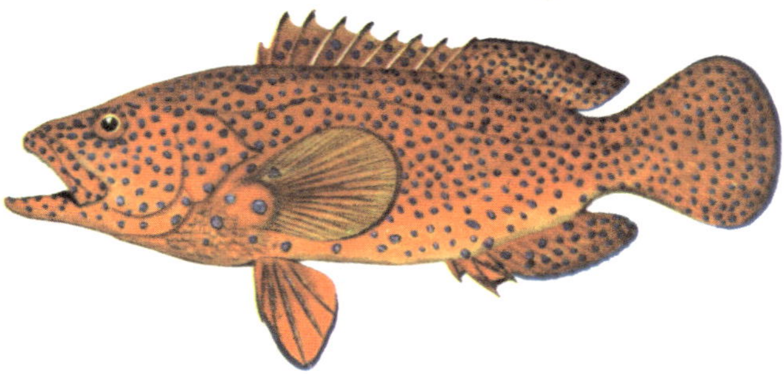

1. Blue-spotted Rock Cod, *Cephalopholis cyanostigma,* 28cm.
2. Freckled Rock Cod. *Cephalopholis sexmaculatus,* 29cm.
3. Coral Trout, *Cephalopholis miniatus,* 25cm.

PLATE 13

1

2

3

1. Yellow-spotted Rock Cod, *Epinephelus areolatus,* 30cm.
2. Black-tipped Rock Cod, *Epinephelus fasciatus,* 25cm.
3. Estuary Rock Cod, *Epinephelus tauvina,* 48cm.

PLATE 14

1

2

3

1. Groper, *Epinephelus lanceolatus,* 53cm.
2. Honeycomb Rock Cod, *Epinephelus merra,* 30cm.
3. Coral Cod or Leopard-fish, *Plectropomus maculatus,* 56cm.

PLATE 15

1

2

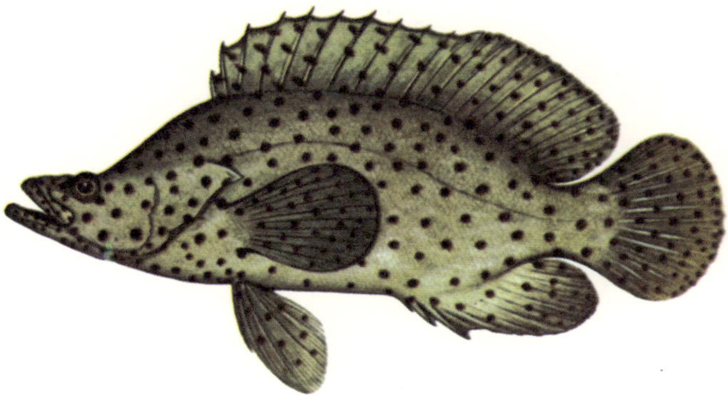

1. White-lined Rock Cod, *Anyperodon leucogrammicus,* 26cm.
2. Hump-backed Rock Cod, *Cromileptes altivelis,* 50cm.

PLATE 16

1

2

3

1. Crescent Perch, *Terapon jarbua*, 26cm.
2. Trumpeter Perch, *Pelates quadrilineatus*, 9cm.
3. "Dolphin", *Coryphaena hippurus*, 90cm.

PLATE 17

1

2

1. Diamond-fish, *Mønodactylus argenteus,* 13cm.
2. Steelback, *Leptobrama mülleri,* 28cm.

PLATE 18

1 

2 

Chinaman-fish, *Symphorus nematophorus,*
1. juv., 30cm;
2. adult, 68cm.

PLATE 19

1

2

1. Mangrove Jack, *Lutjanus argentimaculatus*, 35cm.
2. Blue-banded Sea Perch, *Lutjanus kasmira*, 28cm.

PLATE 20

1

2

Red Emperor, *Lutjanus sebae,*
1. juv., 28cm;
2. adult, 90cm.

PLATE 21

1

2

3

1. Yellow-banded Hussar, *Lutjanus amabilis,* 30cm.
2. Scarlet Sea Perch, *Lutjanus malabaricus,* juv., 35cm.
3. Scarlet Sea Perch, *Lutjanus malabaricus*, 83cm.

PLATE 22

1

2

3

1. Black-spot Sea Perch, *Lutjanus fulviflamma*, 28cm.
2. Kelp Sea Perch, *Lutjanus bohar*, 71cm.
3. Red-bellied Fusilier, *Caesio erythrogaster*, 25cm.

PLATE 23

1

2

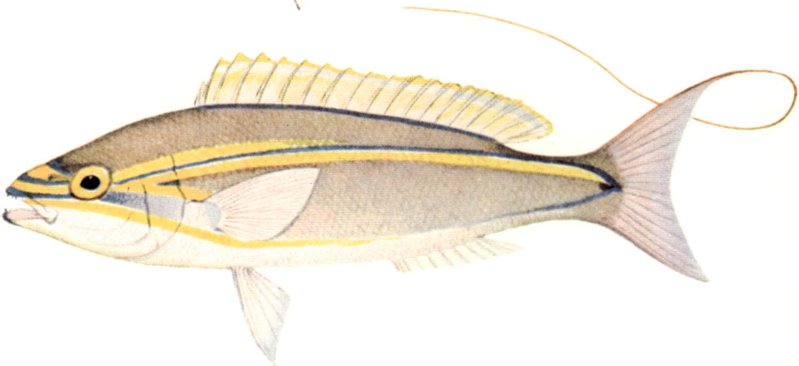

1. Pearly Spine-cheek, *Scolopsis margaritifer*, 23cm.
2. Blue-banded Whiptail, *Pentapodus setosus*, 28cm.

PLATE 24

1 

2 

1. Spotted Javelin-fish, *Pomadasys hasta,* 45cm.
2. Painted Sweet-lips, *Plectorhynchus pictus,* 30cm.

PLATE 25

1

2

1. Yellow-tailed Emperor, *Lethrinus mahsena,* 35cm.
2. Spangled Emperor, *Lethrinus nebulosus,* 30cm.

PLATE 26

1

2

3

1. Sweet-lip Emperor, *Lethrinus chrysostomus*, 43cm.
2. Red-finned Emperor, *Lethrinus nebulosus*, 30cm.
3. Collared Sea Bream, *Gymnocranius audleyi*, 26cm.

PLATE 27

1 

2 

1. Snapper, *Chrysophrys auratus*, 76cm.
2. Long-spined Snapper, *Argyrops spinifer*, 28cm.

PLATE 28

1

2

1. Tarwhine, *Rhabdosargus sarba,* 30cm.
2. Yellowfin or Common Bream, *Acanthopagrus australis,* 19cm.

PLATE 29

1

2

3

1. Pikey Bream, *Acanthopagrus berda,* 35cm.
2. Bearded Jewfish, *Sciaena dussumieri,* 20cm.
3. Great Trevally, *Caranx sexfasciatus,* 30cm.

PLATE 30

1

2

3

1. Turrum, *Carangoides emburyi*, 53cm.
2. Herring Trevally, *Alepes kalla*, 10cm.
3. Fringe-finned Trevally, *Absalom radiatus*, 33cm.

PLATE 31

1

2

1. Golden Trevally, *Gnathanodon speciosus,* juv., 35cm.
2. Black Pomfret, *Parastromateus niger,* juv., 18cm.

PLATE 32

1

2

3

1. Large-mouthed Leather-skin, *Scomberoides lysan,* 71cm.
2. Runner, *Elagatis bipinnulatus,* 58cm.
3. King Amber-Jack, *Seriola lalandi,* 90cm.

PLATE 33

1

2

1. Moon-fish, *Mene maculata,* 19cm.
2. Pig-nosed Pony-fish, *Secutor ruconius,* 6cm.

PLATE 34

1

2

1. Spotted Butter-fish, *Scatophagus argus,* 18cm.
2. Beaked Coral-fish, *Chelmon rostratus,* 13cm.

PLATE 35

1

2

1. Long-bill, *Forcipiger longirostris,* 14cm.
2. Six-spined Butterfly-fish, *Parachaetodon ocellatus,* 11cm.

PLATE 36

1

2

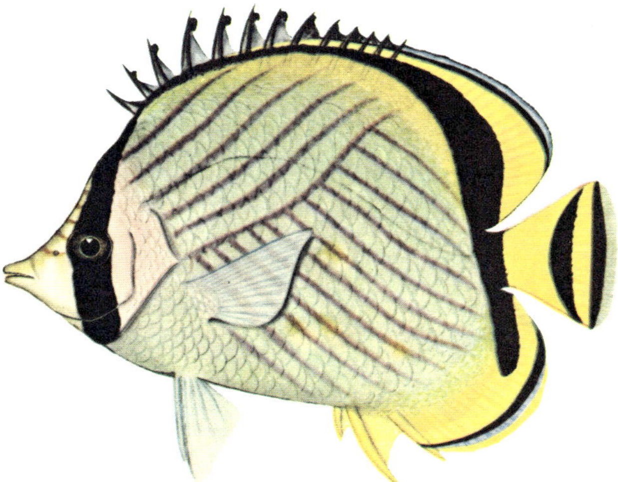

1. Thread-fin Butterfly-fish, *Chaetodon auriga,* 23cm.
2. Criss-cross Butterfly-fish, *Chaetodon vagabundus,* 11cm.

PLATE 37

1

2

1. Lined Butterfly-fish, *Chaetodon lineolatus*, 25cm.
2. Golden-striped Butterfly-fish, *Chaetodon aureofasciatus*, 15cm.

PLATE 38

1

2

1. Right-angled Butterfly-fish,
Chaetodon trifascialis, 18cm.
2. Feather-fin Bull-fish,
Heniochus acuminatus, 10cm.

PLATE 39

1

2

1. Scribbled Angel-fish, *Chaetodontoplus duboulayi,* 20cm.
2. Sickle-fish, *Drepane punctata,* 20cm.

PLATE 40

Batfish, *Platax pinnatus,* juv., 10cm.

Batfish, *Platax pinnatus,* adult, 30cm.

PLATE 42

1

2

3

1. Orange Anemone-fish, *Amphiprion percula,* 5.5cm.
2. Blue Puller, *Chromis caeruleus,* 7.5cm.
3. Jerdon's Damsel-fish, *Pristotis jerdoni,* 12cm.

PLATE 43

1 

2 

1. Harlequin Tusk-fish, *Lienardella fasciatus,* 23cm.
2. Purple Tusk-fish, *Choerodon cephalotes,* 38cm.

PLATE 44

1 

2 

1. Venus Tusk-fish, *Choerodon venustus,* 50cm.
2. Blue Tusk-fish, *Choerodon albigena,* 30cm.

PLATE 45

1 

2 

3 

1. Sling-jaw, *Epibulus insidiator*, 13cm.
2. Black-eyed Thick-lip, *Hemigymnus melapterus*, adult, 29cm.
3. Black-eyed Thick-lip, *Hemigymnus melapterus*, juv., 13cm.

PLATE 46

1

2

1. Clouded Rainbow-fish, *Halichoeres nebulosus,* 8cm.
2. Saddled Rainbow-fish, *Halichoeres margaritaceus,* 10cm.

PLATE 47

1 

2 

3 

1. Moon Wrasse, *Thalassoma lunare,* 20cm.
2. Scarlet-breasted Maori Wrasse, *Cheilinus fasciatus,* 30cm.
3. Hump-headed Maori Wrasse, *Cheilinus undulatus,* 1.45m.

PLATE 48

1

2

3
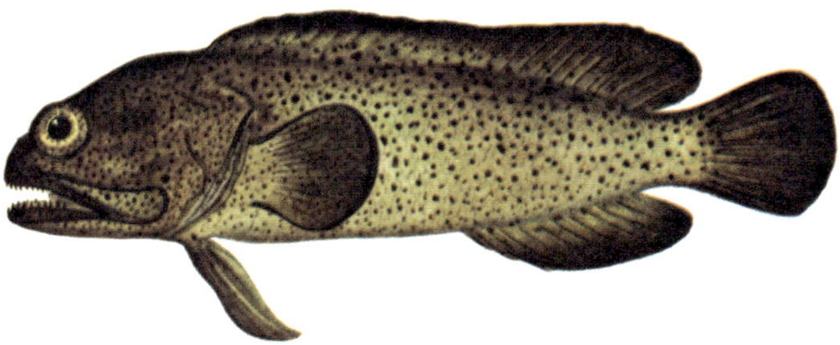

1. Surf Parrot-fish, *Scarus fasciatus,* male 25cm.
2. Surf Parrot-fish, *Scarus fasciatus,* female, 30cm.
3. Speckled Pug, *Opistognathus papuensis,* 24cm.

PLATE 49

1

2

3

1. Bar-faced Weever, *Parapercis nebulosus,* 23cm.
2. Star-gazer, *Ichthyscopus sannio,* 25cm.
3. Five-banded Surgeon-fish, *Acanthurus triostegus,* 18cm.

PLATE 50

1

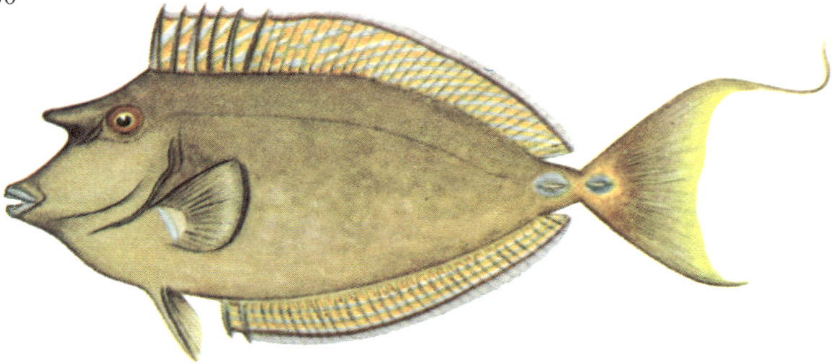

2

3

1. Brown Unicorn-fish, *Naso unicornis,* 30cm.
2. Golden-lined Spinefoot, *Siganus lineatus,* 25cm.
3. Indo-Pacific Sail-fish, *Istiophorus platypterus,* 3m.

1

2

3

4

1. Long-jawed Mackerel, *Rastrelliger kanagurta*, 25cm.
2. Common Mackerel, *Scomber australasicus*, 25cm.
3. Scaleless Tuna, *Gymnosarda unicolor*, 88cm.
4. Mackerel Tuna or Little Tuna, *Euthynnus affinis*, 1m.

PLATE 52

1

2

3

4

1. Watson's Bonito, *Cybiosarda elegans,* 33cm.
2. Pacific Yellow-finned Tuna, *Thunnus albacares,* 1m.
3. Northern Blue-finned Tuna, *Thunnus tonggol,* 1.15m.
4. Australian Bonito, *Sarda australis,* 23cm.

PLATE 53

1

2

3

1. Large-scaled Tunny, *Grammatorycnus bicarinatus*, 1m.
2. Bynoe's Goby, *Amblygobius bynoensis*, 8cm.
3. Fingered Dragonet, *Dactylopus dactylopus*, male, 16cm.

PLATE 54

1

2

3

4

1. Mandarin-fish, *Synchiropus splendidus,* 8cm.
2. Pick-handle Barracuda, *Sphyraena jello,* 76cm.
3. Striped Barracuda, *Sphyraena obtusata,* 25cm.
4. Shark Mullet, *Rhinomugil nasutus,* 18cm.

PLATE 55

1

2

3

1. Diamond-scaled Mullet, *Liza vaigiensis,* 31cm.
2. Bluetail Mullet, *Valamugil seheli,* 43cm.
3. Giant Threadfin, *Eleutheronema tetradactylum,* 40cm.

PLATE 56

1

2

3

1. Sheridan's Threadfin, *Polynemus sheridani,* 45cm.
2. Striped Threadfin, *Polynemus plebeius,* 48cm.
3. Flat Salmon, *Polynemus multiradiatus,* 20cm.

PLATE 57

1

2

1. Marbled Coral Cod, *Scorpaena bynoensis,* 7.5cm.
2. Bullrout, *Notesthes robusta,* 25cm.

PLATE 58

Red Fire-fish, *Pterois volitans,* 18cm.

PLATE 59

1

2

1. Purple Flying Gurnard, *Dactyloptena orientalis,* 28cm.
2. Dorsal view with pectoral fins extended.

PLATE 60

1

2

3

1. Long-finned Gurnard
Lepidotrigla calodactylus, 13.5cm.
2. Extended pectoral fin
(Long-finned Gurnard).
3. Rufous Flying Gurnard,
Dactyloptena papilio, 18cm.

1 

2 

3 

1. Fringe-eyed Flathead, *Platycephalus nematophthalmus,* 25cm.
2. Dwarf Flathead, *Elates ransonetti,* 19.5cm.
3. Long-tailed Sea Dragon, *Parapegasus natans,* 10cm.

PLATE 62

1 

2 

1. Queensland Halibut of Adalah, *Psettodes erumei,* 38cm.
2. Large-toothed Flounder, *Pseudorhombus arsius,* 27cm.

1 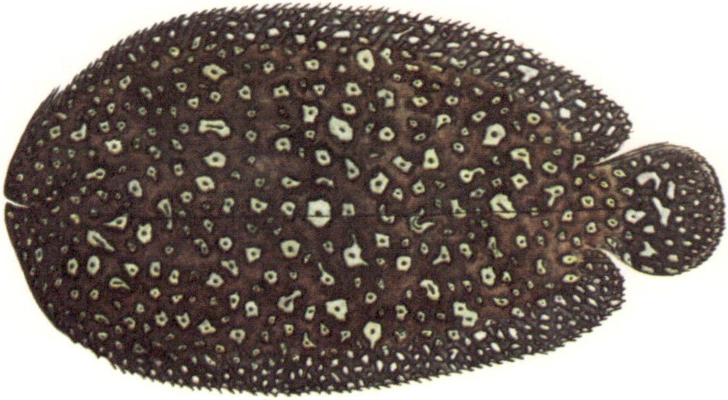

2 

1. Peacock Sole, *Pardachirus pavoninus,* 25cm.
2. Sharp-headed Sole, *Phyllichthys sclerolepis,* 13cm.

PLATE 64

1

2

3

1. Two-lined Tongue Sole, *Cynoglossus bilineatus,* 20cm.
2. Slender Sucking-fish, *Echeneis naucrates,* 58cm.
3. Black-finned Triple-spine, *Triacanthus biaculeatus,* 18cm.

PLATE 65

1 

2 

1. Big-spotted Trigger-fish, *Balistoides conspicillum,* 27cm.
2. Flat-tailed Trigger-fish, *Abalistes stellaris,* 25cm.

PLATE 66

1

2

1. Red-lined Trigger-fish, *Balistapus undulatus,* 35cm.
2. Spotted Trigger-fish, *Canthidermis rotundatus,* 10cm.

PLATE 67

1

2

1. Beardie or Tape-fish, *Anacanthus barbatus,* 16cm.
2. Prickly Leather-jacket, *Chaetoderma penicilligera,* 20cm.

PLATE 68

1

2

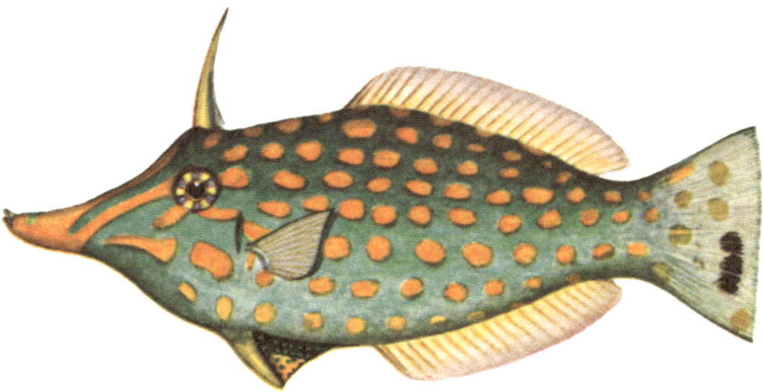

1. Fan-bellied Leather-jacket, *Monacanthus chinensis,* 23cm.
2. Beaked Leather-jacket, *Oxymonacanthus longirostris,* 11cm.

PLATE 69

1

2

3

1. Figured Leather-jacket, *Alutera scripta*, 31cm.
2. Long-nosed Box-fish, *Rhynchostracion nasus*, 35cm.
3. Blue-spotted Box-fish, *Ostracion cubicus*, 15cm.

PLATE 70

1

2

3

1. Cow-fish, *Lactoria cornutus,* 20cm.
2. Banded Toado, *Torquigener pleurostictus,* 12cm.
3. Stars and Stripes Toado, *Arothron hispidus,* 29cm.

PLATE 71

1 

2 

3 

1. A Toado, *Arothron stellatus,* 56cm.
2. Marbled Toado, *Chelonodon patoca,* 18cm.
3. Porcupine-fish, *Tragulichthys jaculiferus,* 23.5cm.

PLATE 72

Marbled Angler, *Histrio histrio,* 6cm.

PLATE 73

A

B

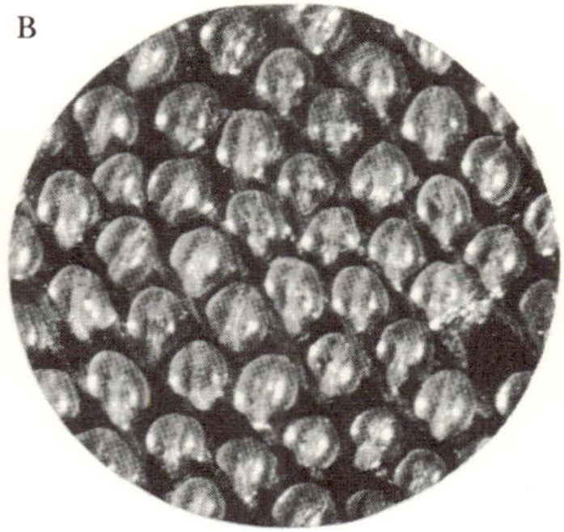

C

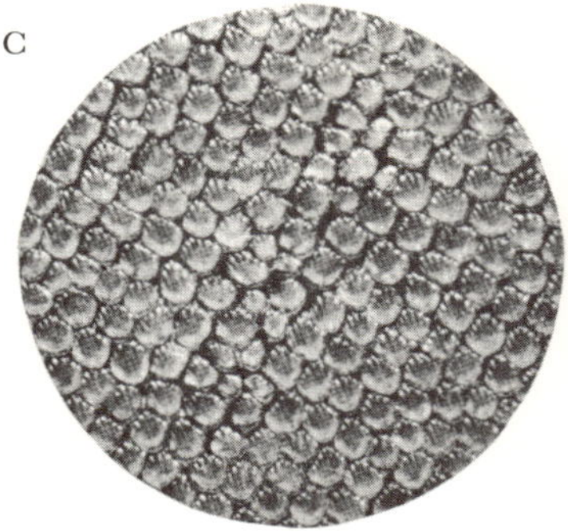

D

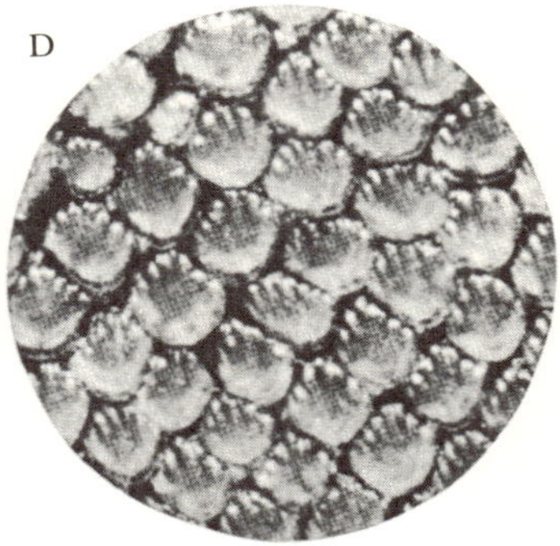

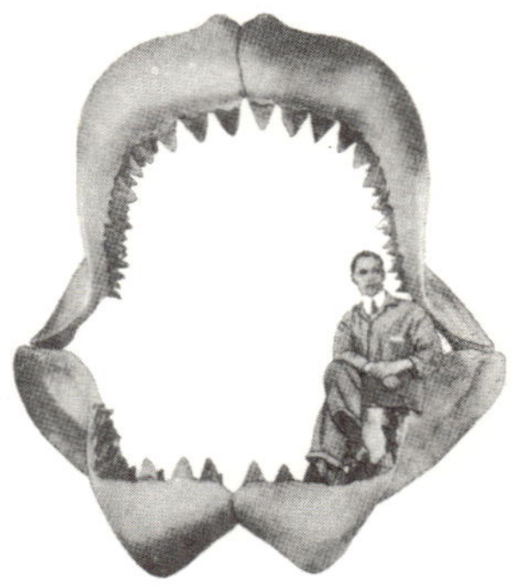

Sections of shark's skins, highly magnified, showing dermal denticles of shagreen of:
A, Wobbegong (x9);
B, Tiger Shark (x9);
C, Whaler Shark (x9);
D, Whaler Shark (x18).
Photos: T.C. Roughley.

Reconstructed jaws and teeth of a giant white shark, thought to be only recently extinct. *American Museum Journal.*

PLATE 74

1

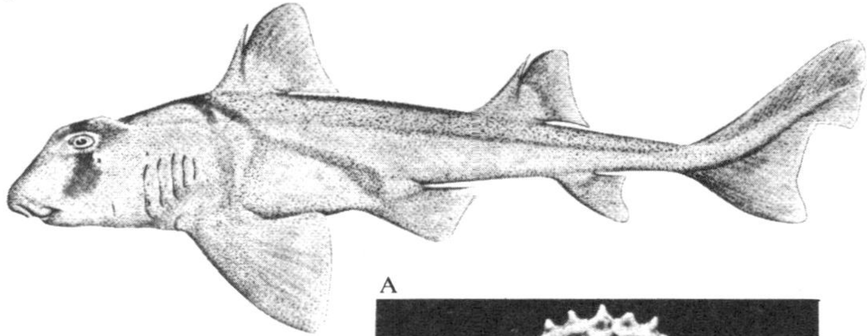

A

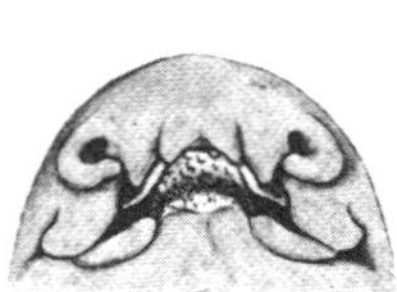

2

1. Port Jackson Shark, *Heterodontus portusjacksoni,* female, 56cm.
   A, jaw showing enlarged lateral teeth without keels.
2. Crested Port Jackson Shark, *Heterodontus galeatus,* with detail of mouthparts.

1

2

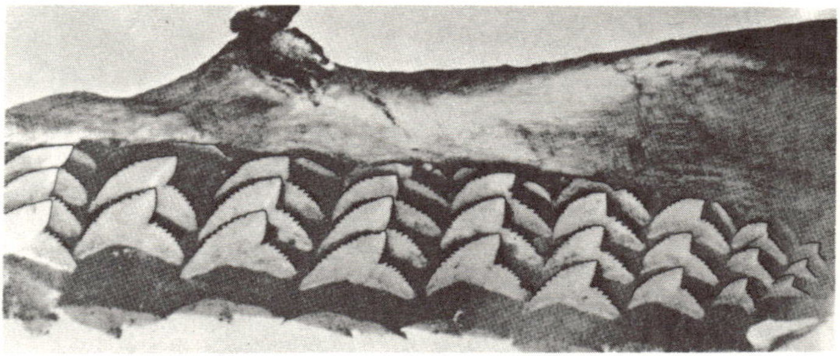

3

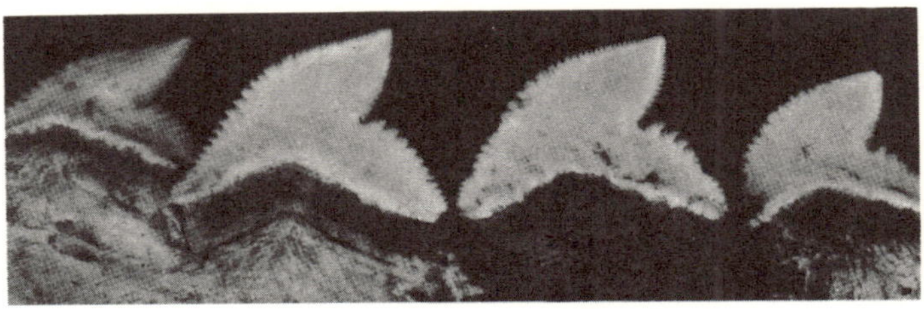

4

1. Tiger Shark, *Galeocerdo cuvier,* about 2.25m.
2. A portion of the dried jaws with flesh removed, showing the rows of unerupted teeth lying against the jaw.
3. Front row of functional teeth.
4. Weasel Shark, *Hemigaleus microstoma,* about 72cm.

PLATE 76

1

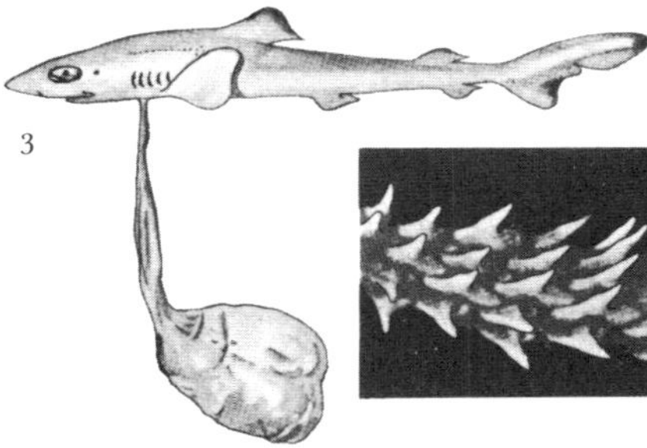

3

2

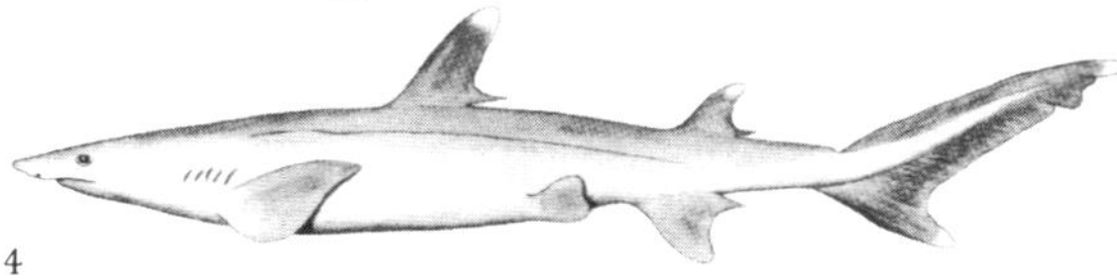

4

5

1. School Shark, *Galeorhinus australis,* 1.5m.
2. Section of jaw (with teeth).
3. The embryo.
4. White-tip Shark, *Triaenodon obesus,* young female, 71cm.
5. Blackspot Shark, *Carcharhinus sealei,* 92cm.

PLATE 77

1. Dusky Shark, *Carcharhinus obscurus,* 88cm. Also (from left) upper and lower tooth of same species, scale from shoulder, and under-surface view of head.
2. Bronze Whaler, *Carcharhinus brachyura,* 3m.
3. Graceful Shark, *Carcharhinus amblyrhynchoides,* female, 60cm.
4. Black-tip Shark, *Carcharhinus melanopterus.*
5. Estuary Shark, *Carcharhinus leucas,* 45cm.
6. Little Blue Shark, *Scoliodon laticaudus,* 28cm.

PLATE 78

1

2

3

4

5

6

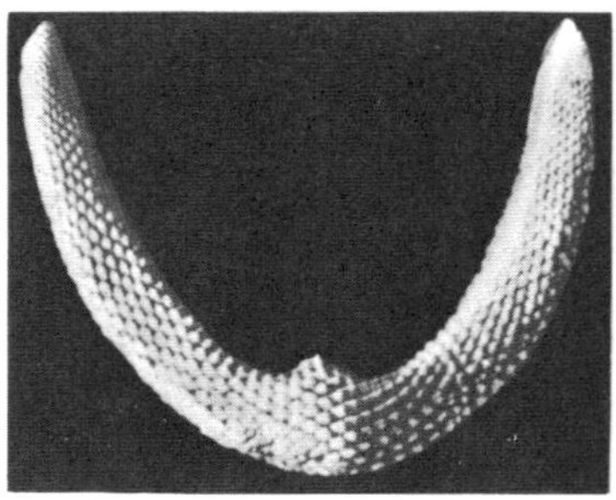

1. Smooth-fanged Shark, *Carcharhinus brevipinna,* 80cm.
2. Lemon Shark, *Negaprion acutidens,* female, 65cm.
3. Taylor's Shark, *Rhizoprionodon taylori,* 83cm.
4. Slender Dog Shark, *Loxodon macrorhinus,* 71cm.
5. Gummy Shark, *Mustelus antarcticus,* 90cm.
6. Jaw showing pavement-like teeth (Gummy Shark).

PLATE 79

1

2

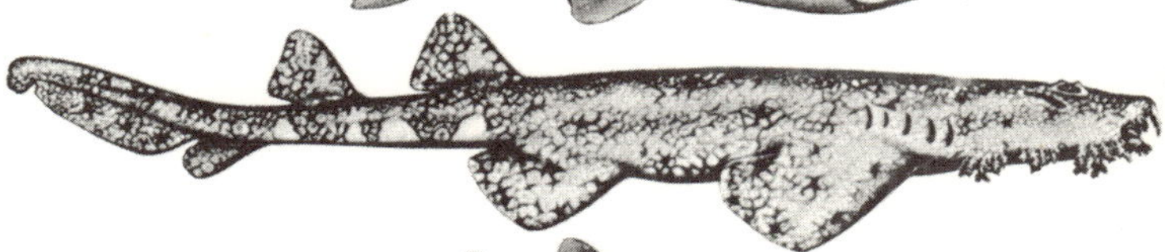

3

4

5

6

1. Hammer-head Shark, *Sphyrna lewini,* female, 58cm. Also shown, under-side of head and teeth.
2. Tawny Shark, *Nebrius concolor,* 90cm. See also PLATE 81.
3. Tasselled Wobbegong, *Orectolobus ogilbyi,* 1.26m.
4. Spotted Wobbegong, *Orectolobus maculatus,* foetus, 21cm.
5. Banded Wobbegong, *Orectolobus ornatus,* foetus, 19cm.
6. Section of jaw of adult Banded Wobbegong showing fang-like teeth.

PLATE 80

1. Northern Wobbegong, *Orectolobus wardi.*
2. Blind Shark, *Brachaelurus waddi,* 51cm.
3. Blue-grey Cat Shark, *Brachaelurus colcloughi,* 45cm.
4. Epaulette Shark, *Hemiscyllium ocellatum,* 40cm, and section of mouth.
5. Speckled Cat Shark, *Hemiscyllium trispeculare,* 56cm.

PLATE 81

1

2

3

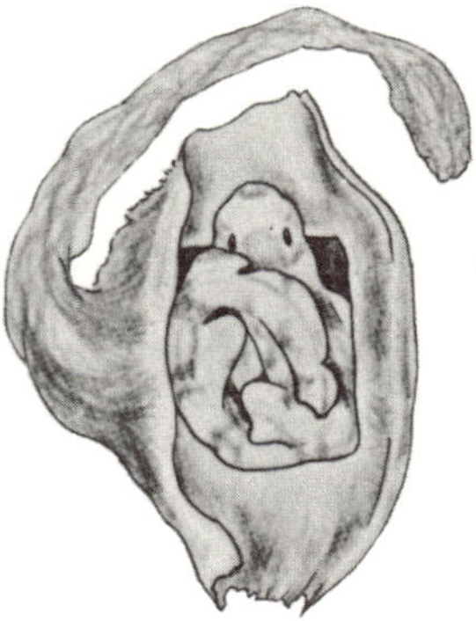

4

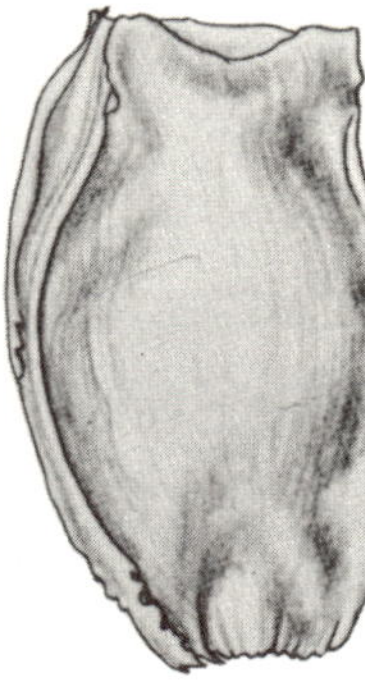

5

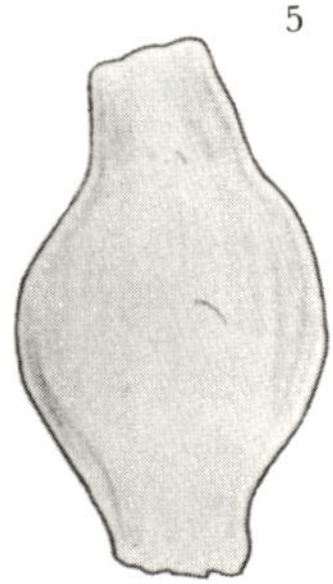

6

7

1. Brown-banded Cat Shark, *Chiloscyllium punctatum,* juv., 38cm.
2. Brown-banded Cat Shark, adult, 53cm.
3. Brown-banded Cat Shark egg.
4. Zebra Shark egg.
5. The egg of the Tawny Shark, *Nebrius concolor* (see PLATE 79).
6. Zebra Shark, *Stegostoma varium,* juv., 35cm.
7. Zebra Shark, *Stegostoma varium,* adult, 2.10m.

PLATE 82

1. Thresher Shark, *Alopias vulpinus,* 3.05m.
2. Grey Nurse, *Odontaspis taurus,* 3m.
3. Blue Pointer or Mako, *Isurus oxyrinchus.*

1

2

1. White Pointer, *Carcharodon carcharias,* about 4m.
2. Section of jaws shows large upright serrated teeth complete as in a live specimen.

PLATE 84

1

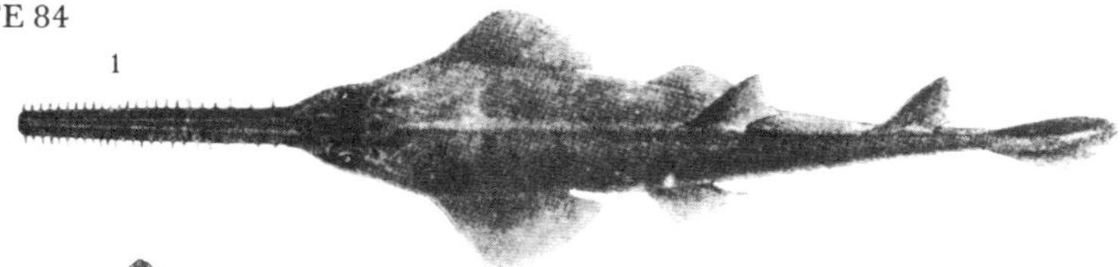

2

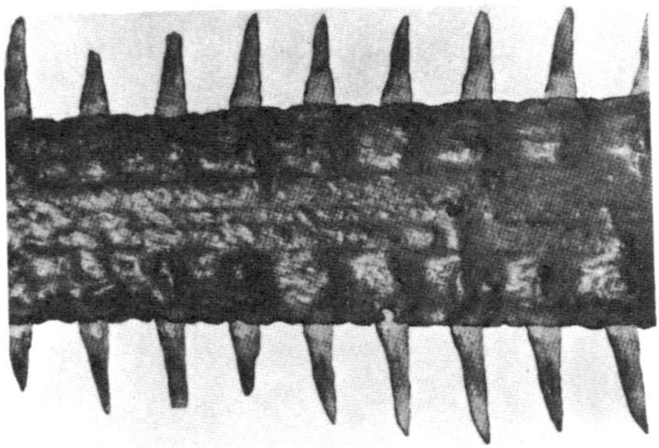

5

3

4

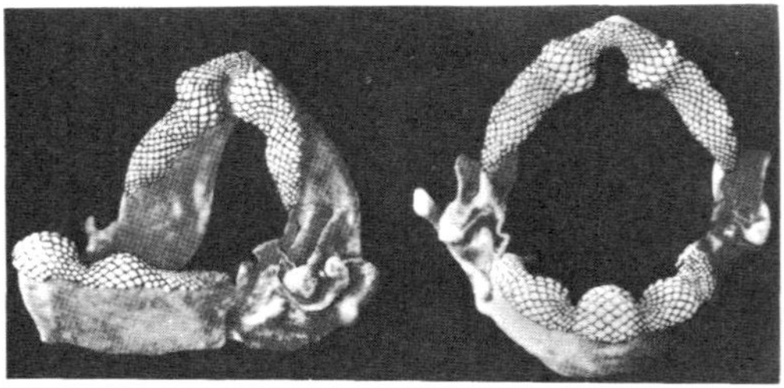

1. Narrow-snouted Saw-fish, *Pristis zijsron,* 86cm.
2. Section of snout or rostral saw of a much larger specimen, probably 38 to 45cm long (full length of this jaw was 1.40m).
3. Shark Ray, *Rhina ancylostomus,* 2m.
4. Two views of the Shark Ray's teeth and jaws. Each tooth has a rippled surface.
5. White-spotted Ray, *Rhynchobatus djiddensis,* 1.25m.

PLATE 85

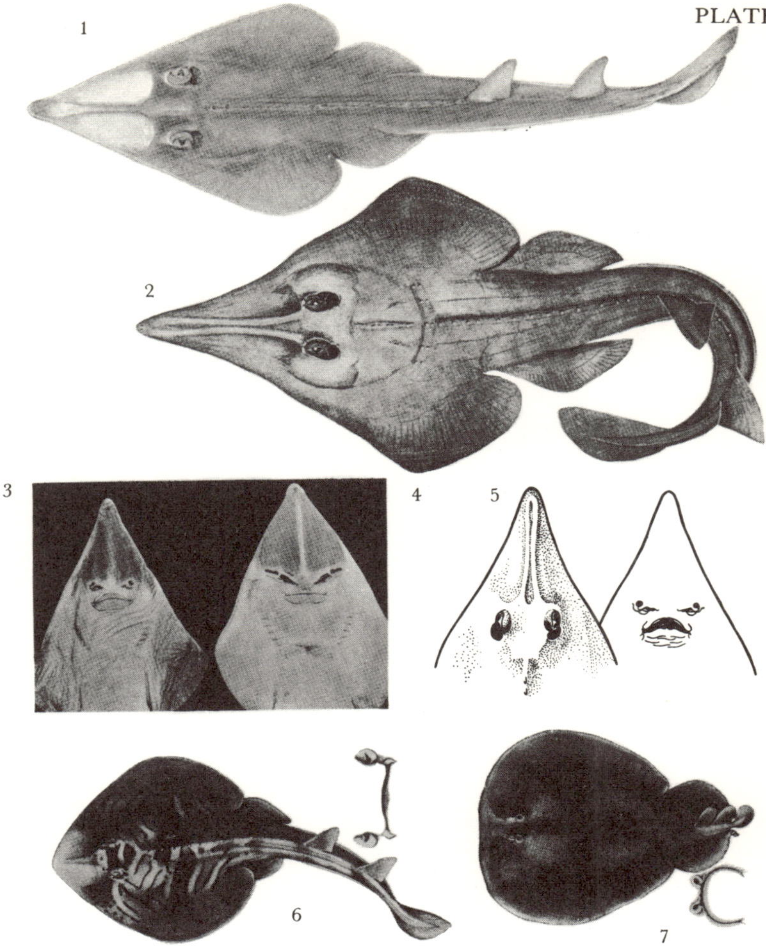

1. Common Shovel-nose Ray, *Rhinobatos batillum,* 48cm.
2. Banks' Shovel-nose Ray, *Aptychotrema rostrata,* 56cm.
3. Under-surface of head (Banks' Shovel-nose Ray).
4. Under-surface of head (Common Shovel-nose Ray).
5. Bougainville's Shovel-nose Ray, *Aptychotrema bougainvillii,* upper and lower surfaces of head of an 81cm male.
6. Fiddler Ray, *Trygonorrhina fasciata,* 38cm.
7. Numb-fish, *Hypnos monopterygium,* 35cm.

PLATE 86

1

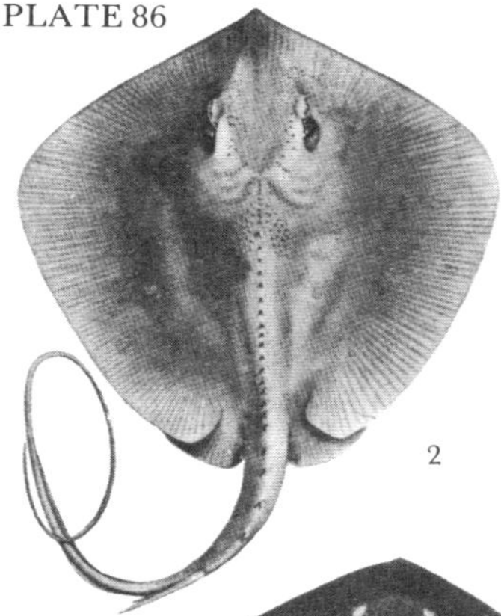

2

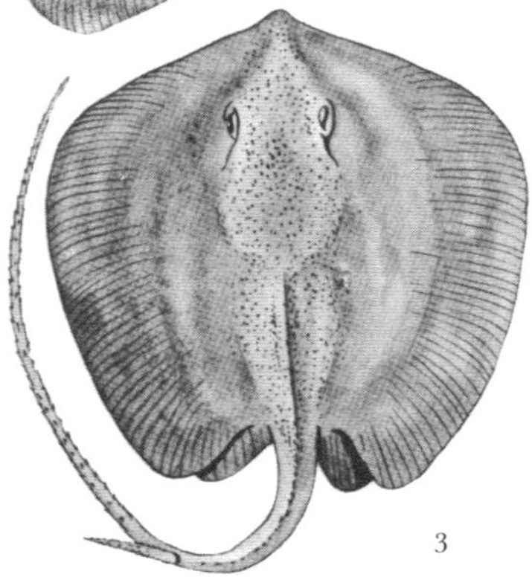

3

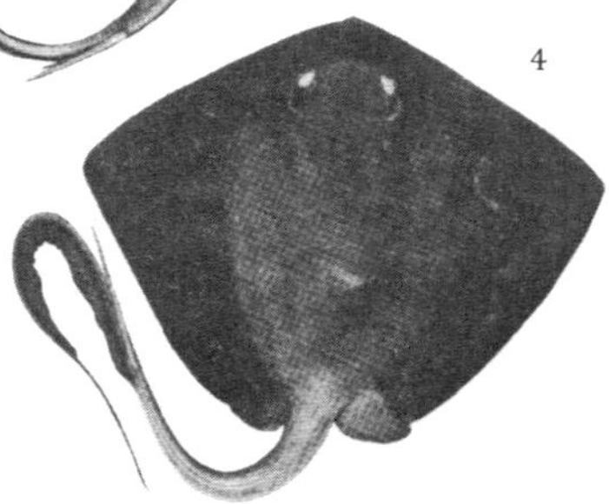

4

5

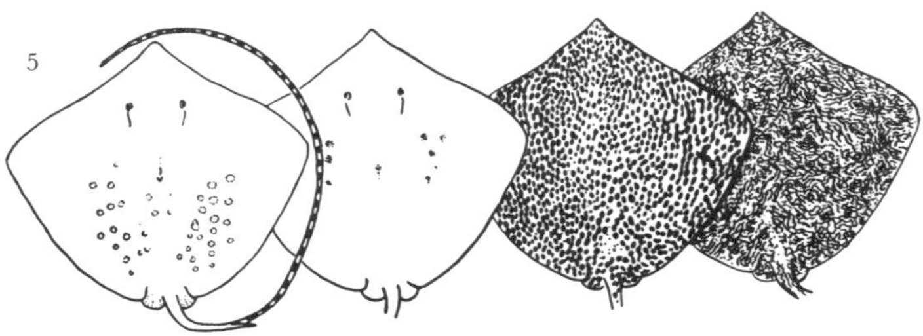

1. Common Skate, *Raja australis,* 48cm.
2. Estuary Stingaree, *Dasyatis fluviorum,* 28cm.
3. Mangrove Ray, *Himantura granulata,* dorsal surface of female, width 33cm.
4. Cow-tail or Fantail Ray, *Dasyatis sephen,* width about 50cm.
5. Variation of pattern of Coach-whip Ray, *Himantura uarnak*

PLATE 87

1

2

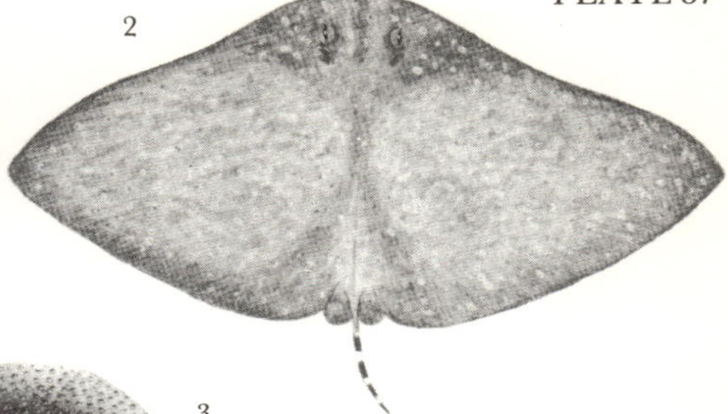

3

4

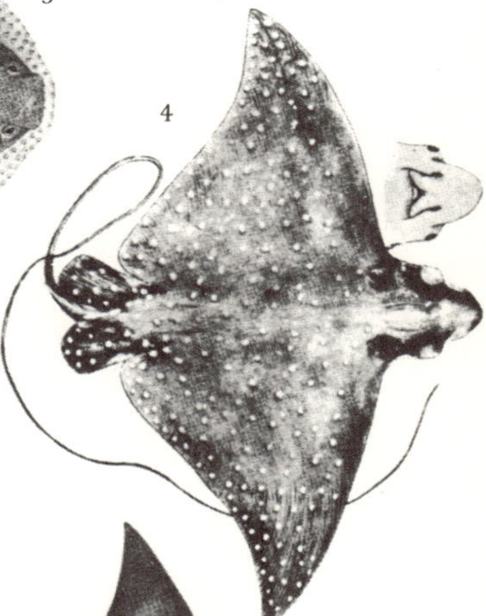

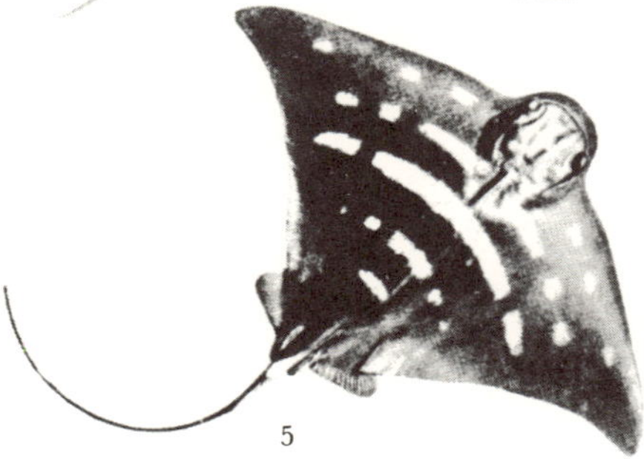

5

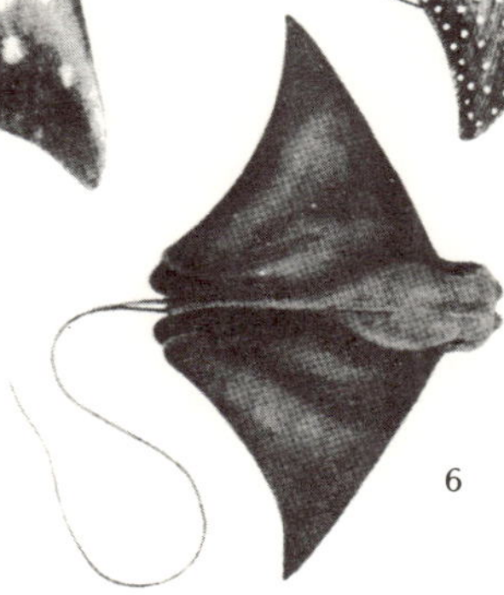

6

1. Common Stingaree, *Urolophus testaceus,* female, width 24cm.
2. Rat-tailed Ray, *Gymnura australis,* width 30cm.
3. Thorny Ray, *Urogymnus asperrimus,* width 60cm.
4. Spotted Eagle Ray, *Aetobatus narinari,* width 56cm.
5. Eagle Ray, *Myliobatus australis,* width 1.25m.
6. Cow-nose Ray, *Rhinoptera neglecta.*

PLATE 88

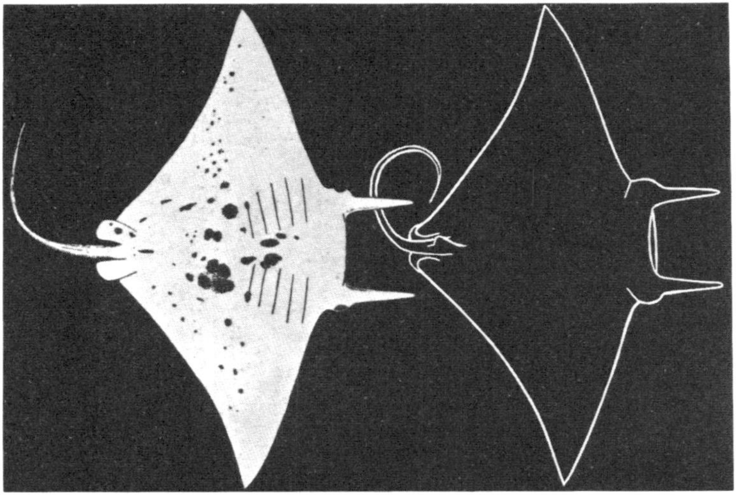

1

2 3

4

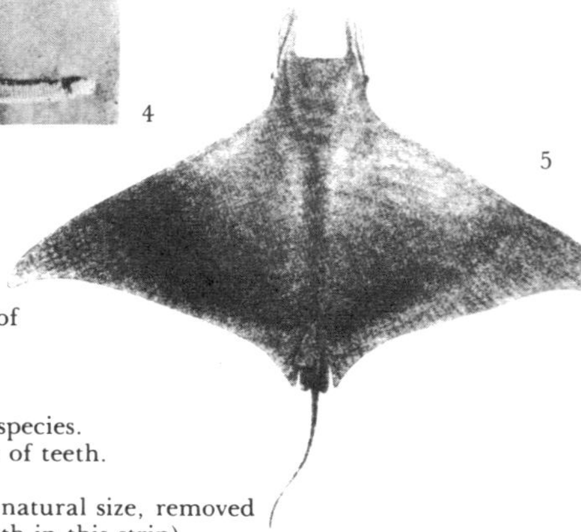

5

1. Australian Devil Ray, *Manta alfredi* (shown one-fiftieth natural size). Outline (right) shows slit-like spiracles behind eyes, also small dorsal fin near root of spineless tail. Ventral surface (left) is white with irregular dark blotches, and the five pairs of gill-slits are plainly seen.
Photographs (2-4) show teeth of one of this species.
2. Skin of lower lip reflected to expose some of teeth.
3. Some of teeth enlarged.
4. Entire dental strip, less than one-quarter natural size, removed from lower jaw (there are 1,500 to 2,000 teeth in this strip).
5. Pygmy Devil Ray, *Mobula diabolus,* width 86cm.

PLATE 89

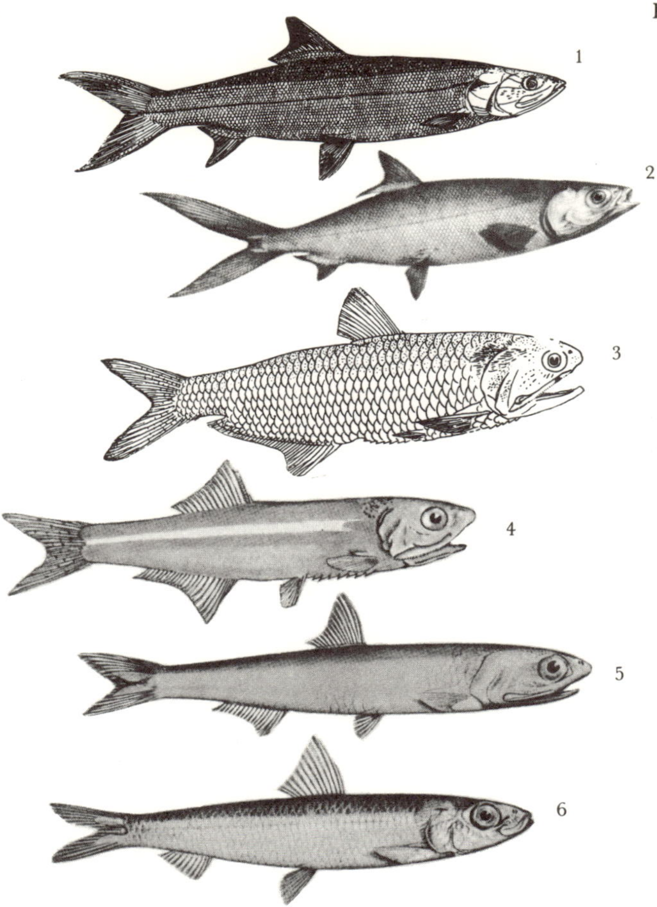

1. Giant Herring, *Elops hawaiiensis,* 33cm.
2. Salmon Herring, *Chanos chanos,* 25cm.
3. Estuary Anchovy, *Thrissina baelama,* 13cm.
4. Gulf Anchovy, *Stolephorus carpentariae,* 5cm.
5. Australian Anchovy, *Engraulis australis,* 11cm.
6. Blue Sprat, *Spratelloides delicatulus,* 6cm.

PLATE 90

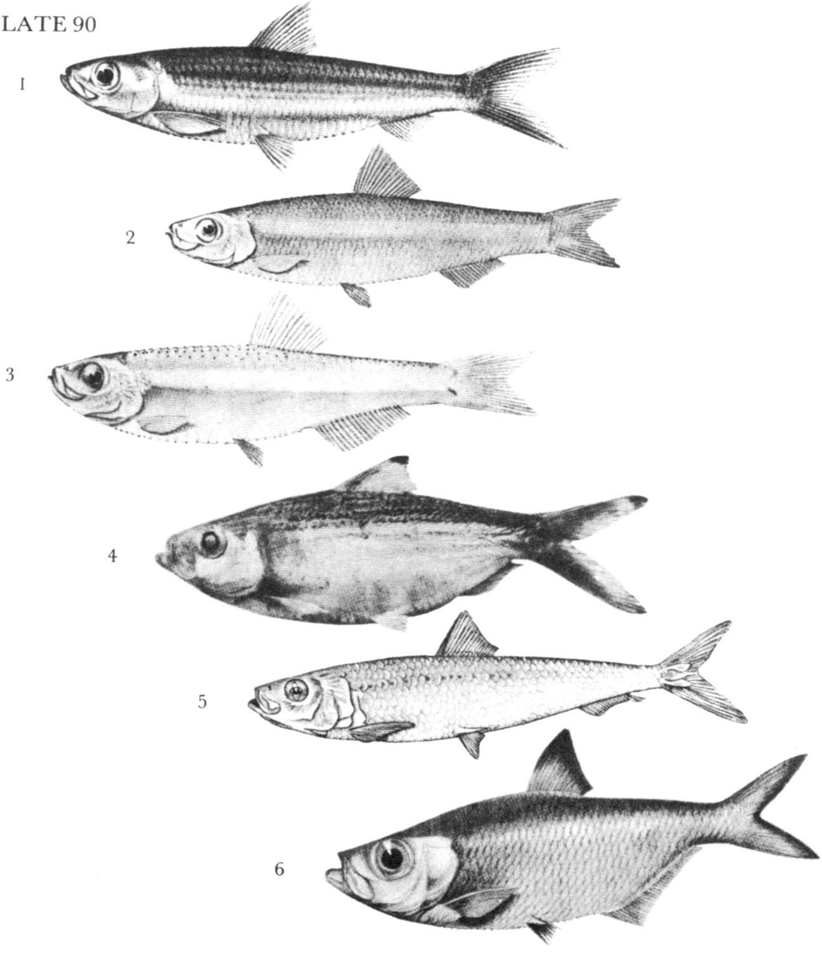

1. Freshwater Herring, *Potamalosa richmondia,* 19cm.
2. Sandy Sprat, *Hyperlophus vittatus,* 8cm.
3. Glassy Sprat, *Hyperlophus translucidus,* 6cm.
4. Southern Herring, *Harengula castelnaui.*
5. Dakin's Pilchard, *Sardinella sirm,* 18.5cm.
6. Ditchelee Herring, *Pellona ditchela,* 17cm.

PLATE 91

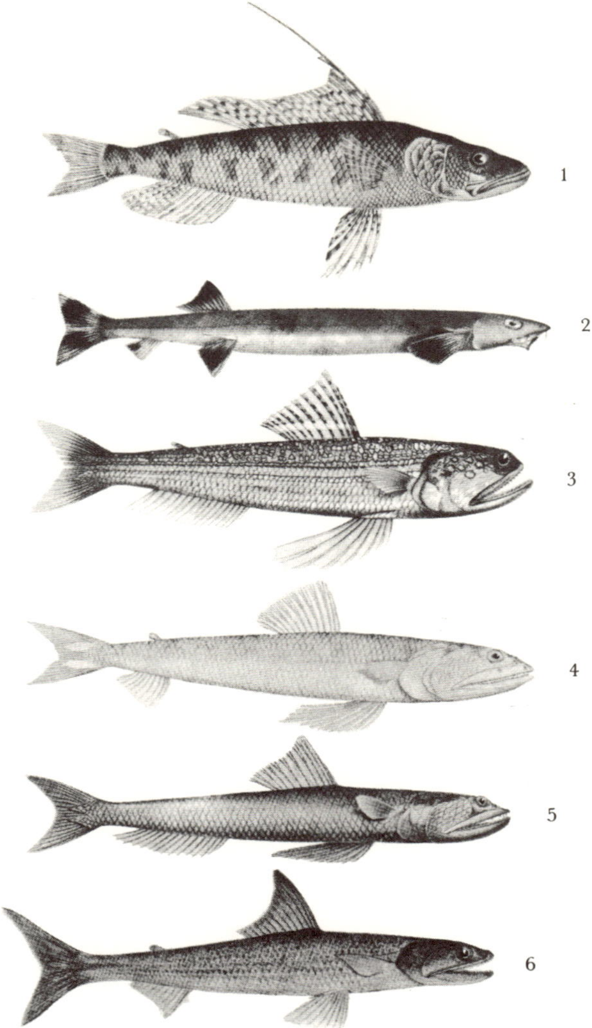

1. Sergeant Baker, *Aulopus purpurissatus,* male, 53cm.
2. Rat-fish, *Gonorhynchus greyi.*
3. Painted Grinner, *Trachinocephalus myops,* 30cm.
4. Hoult's Grinner, *Synodus houlti,* 20cm.
5. Spear-toothed Grinner, *Synodus sageneus,* 26cm.
6. Yellow-banded Grinner, *Saurida tumbil,* about 30cm.

PLATE 92

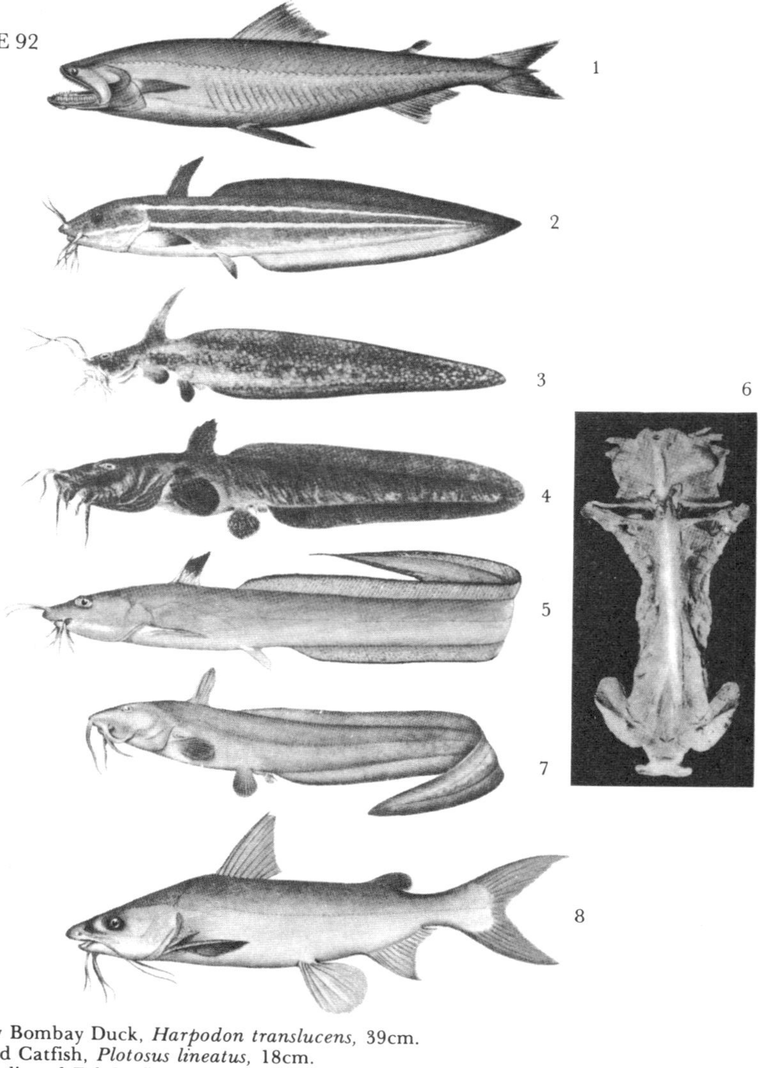

1. Glassy Bombay Duck, *Harpodon translucens,* 39cm.
2. Striped Catfish, *Plotosus lineatus,* 18cm.
3. White-lipped Eel Catfish, *Paraplotosus albilabris,* 23cm.
4. Estuary Catfish, *Cnidoglanis macrocephalus,* 50cm.
5. Nettle-fish, *Euristhmus nudiceps,* 26cm.
6. Skull of a catfish (not same species) showing "crucifix" on under-surface. (Photo from Brisbane *Telegraph*.)
7. Long-tailed Catfish, *Euristhmus lepturus,* 36cm.
8. Giant Salmon Catfish, *Arius thalassina,* 38cm.

PLATE 93

1

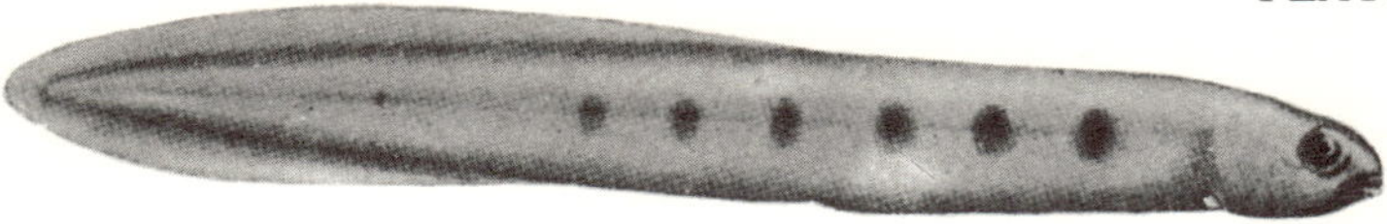

2

3

4

5

1. Red-banded Pygmy Eel or Shore Eel, *Alabes dorsalis*.
2. Long-finned Freshwater Eel, *Anguilla reinhardtii*.
3. Pike Eel, *Muraenesox bagio*, 53cm.
4. Australian Conger Eel, *Conger verreauxi*.
5. Clouded Reef Eel, *Echidna nebulosa*.

PLATE 94

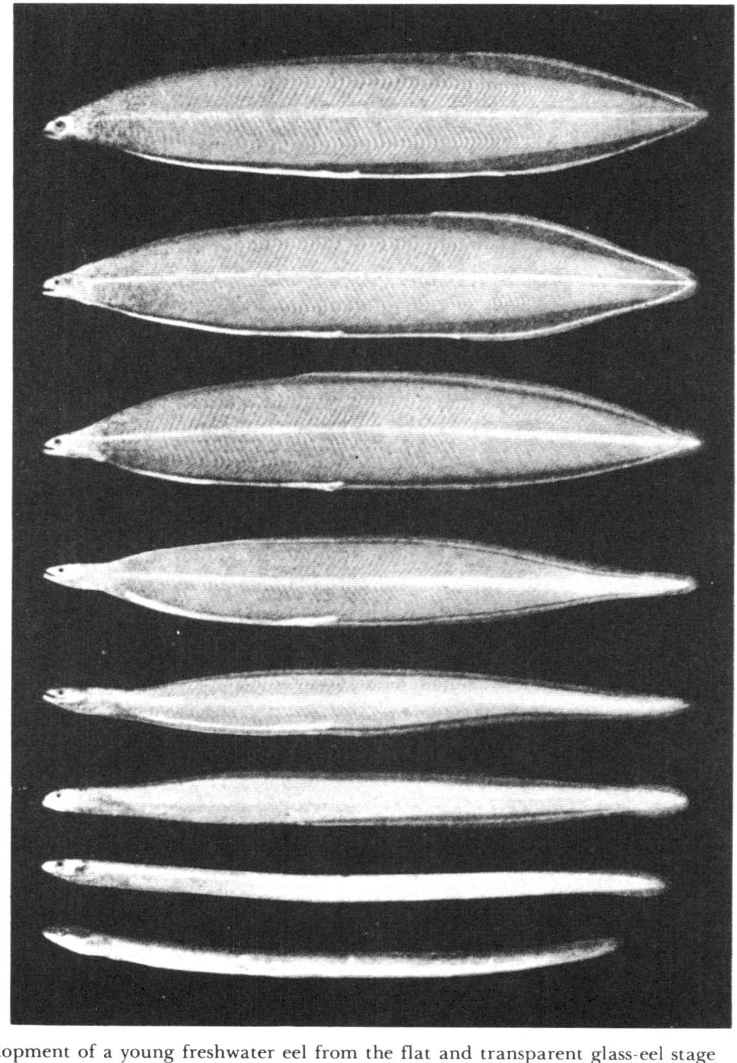

Development of a young freshwater eel from the flat and transparent glass-eel stage to the cylindrical elver or young eel. After Schmidt.

PLATE 95

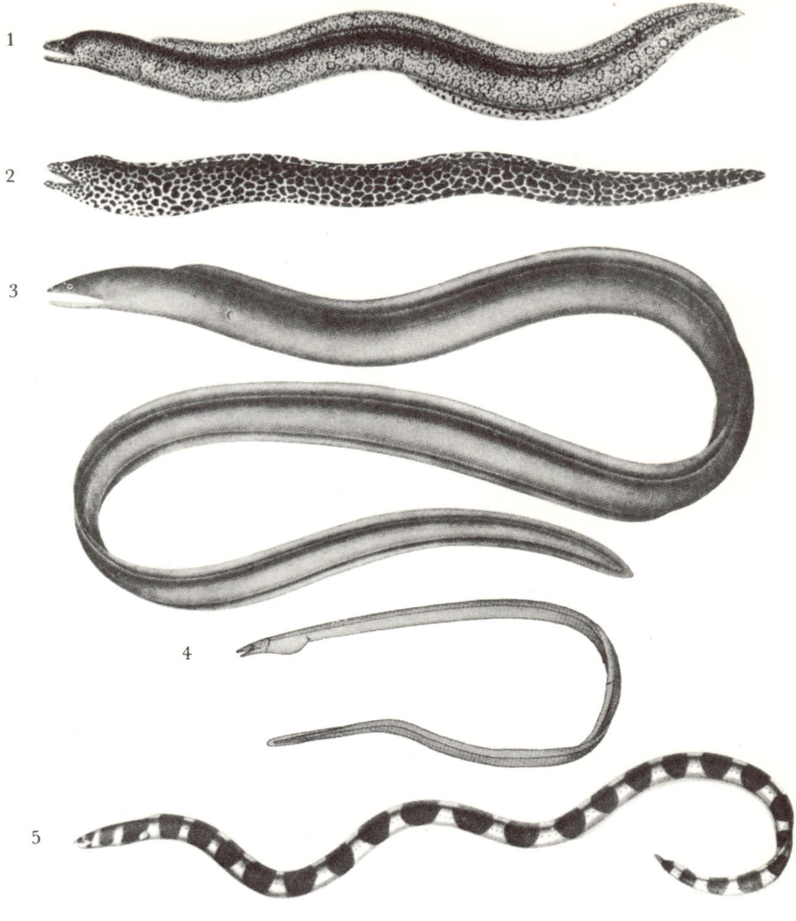

1. Painted Reef Eel, *Gymnothorax pictus,* 35cm.
2. Tessellated Reef Eel, *Gymnothorax favagineus,* 71cm.
3. Long-tailed Eel, *Evenchelys macrurus,* 1.25m.
4. Iredale's Worm Eel, *Muraenichthys iredalei,* 16cm.
5. Half-banded Snake Eel, *Leiuranus semicinctus,* 48cm.

PLATE 96

1

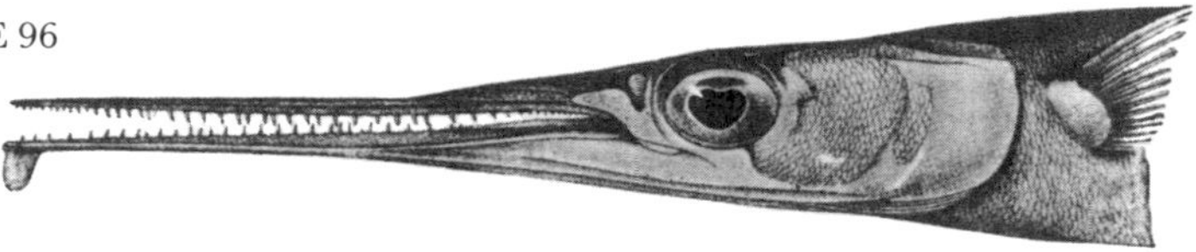

2

3

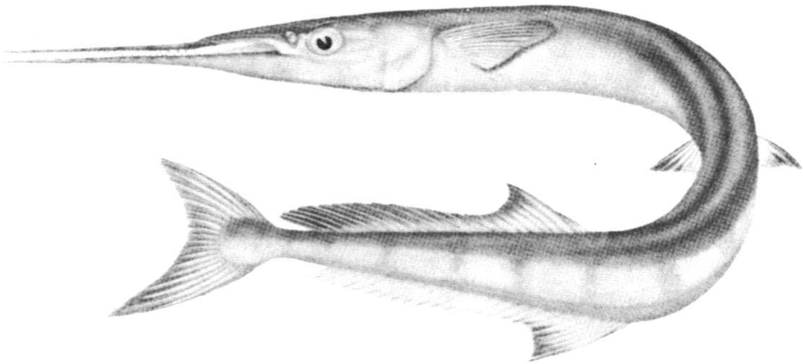

4

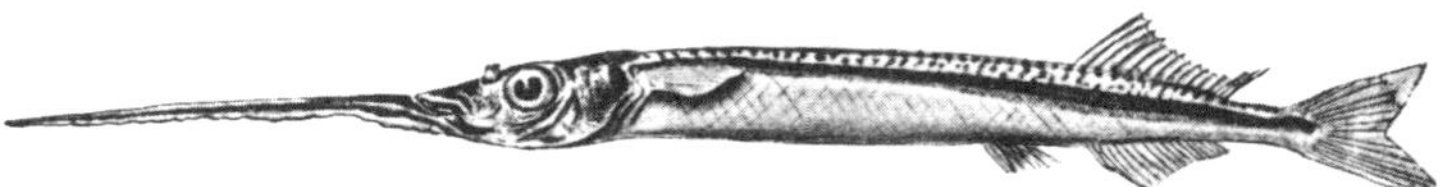

5

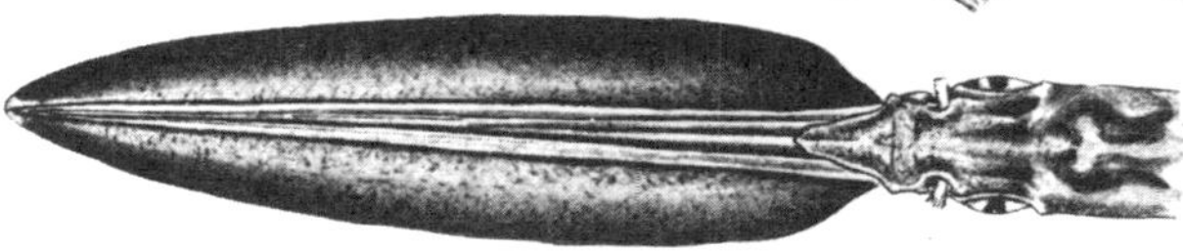

6

1. Black-finned Long Tom, *Tylosurus acus.*
2. Stout Long Tom, *Lhotskia gavialoides.*
3. Barred Long Tom, *Ablennes hians,* 60cm.
4. Long-jawed Garfish, *Rhynchorhampus georgii,* 21cm.
5. Detail of upper surface of head (Long-jawed Garfish).
6. Short-nosed Garfish, *Hyporhamphus quoyi,* 30cm.

PLATE 97

1

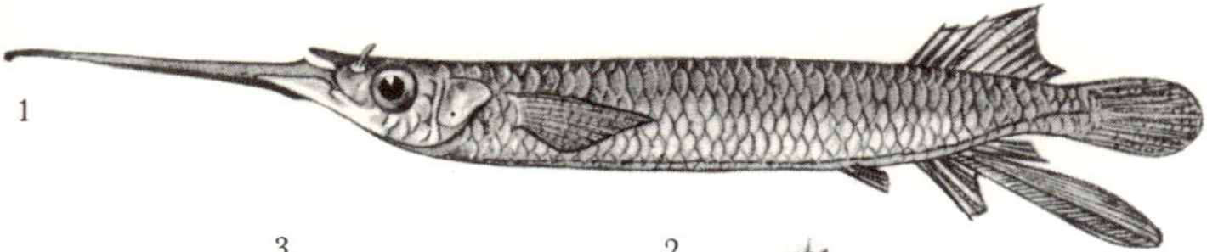

3

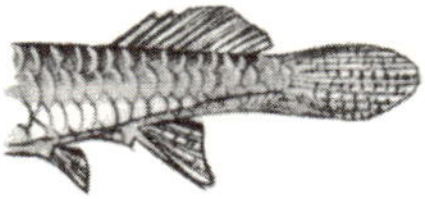

2

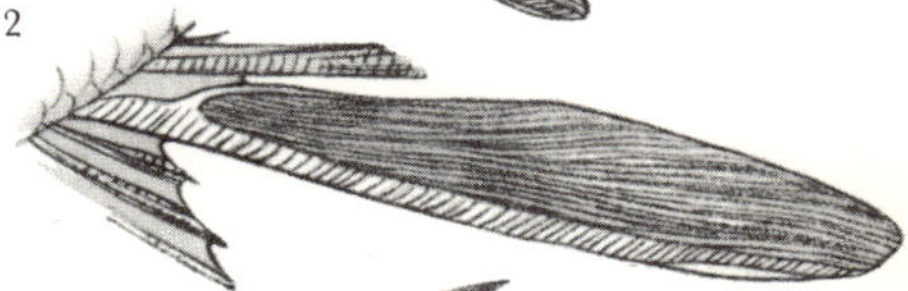

4

5

6

7

1. Spoon-fin Garfish, *Zenarchopterus dispar*, male.
2. Detail of anal fin (magnified), showing enlarged and thickened sixth and seventh ray.
3. Tail end of female (See also PLATE 6.)
4. Long-finned Garfish, *Euleptorhamphus viridis*, 43cm.
5. Short-finned Flying-fish, *Parexocoetus brachypterus*, 11cm.
6. Common Flying-fish, *Exocoetus volitans*, 20cm.
7. Flying-fish, *Hirundichthys oxycephalus*, 20cm.

PLATE 98

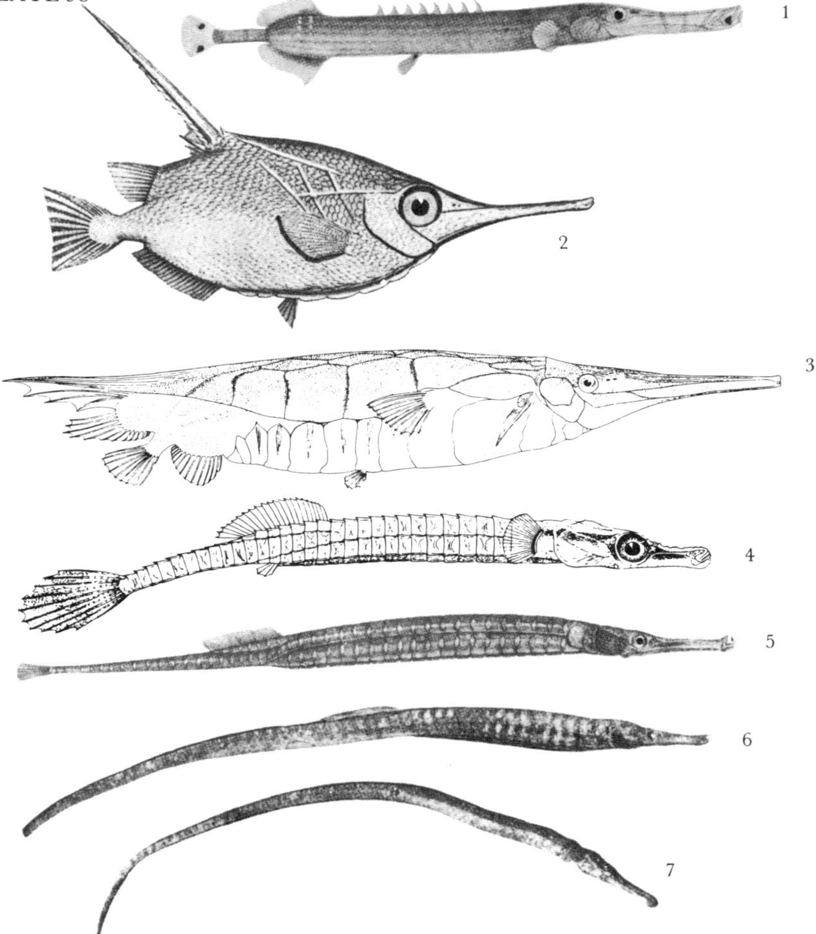

1. Spiny-back Trumpet-fish, *Aulostomus chinensis.*
2. Long-spined Bellows-fish, *Macroramphosus elevatus,* 12cm.
3. Razor-fish, *Centriscus cristatus,* 23cm.
4. Blue-striped Pipe-fish, *Doryhamphus melanopleura,* about 6cm.
5. Short-tailed Pipe-fish, *Microphis manadensis.*
6. Black-chinned Pipe-fish, *Syngnathus altirostris,* male.
7. Black-chinned Pipe-fish, female.

PLATE 99

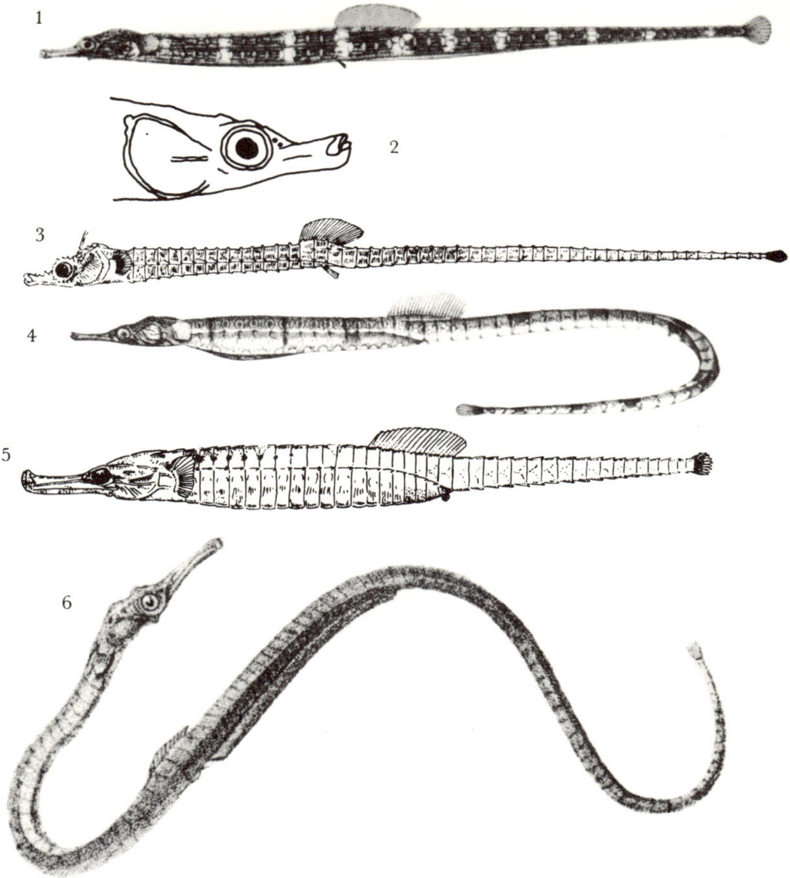

1. Messmate Pipe-fish, *Corythoichthys intestinalis.*
2. Head of Thorn-tailed Pipe-fish, *Micrognathus brevirostris.*
3. Gray's Pipe-fish, *Halicampus grayi,* 9cm.
4. Banded Pipe-fish, *Yozia tigris,* 28cm.
5. Pig Pipe-fish, *Choeroichthys brachysoma,* 5cm.
6. Spotted Pipe-fish, *Yozia bicoarctata,* 28cm.

PLATE 100

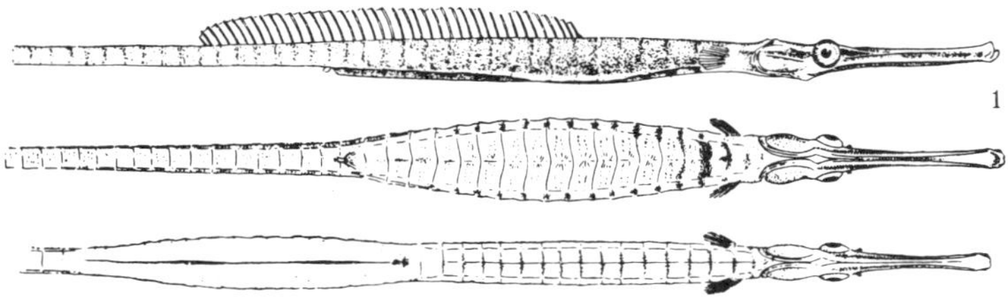

1

2

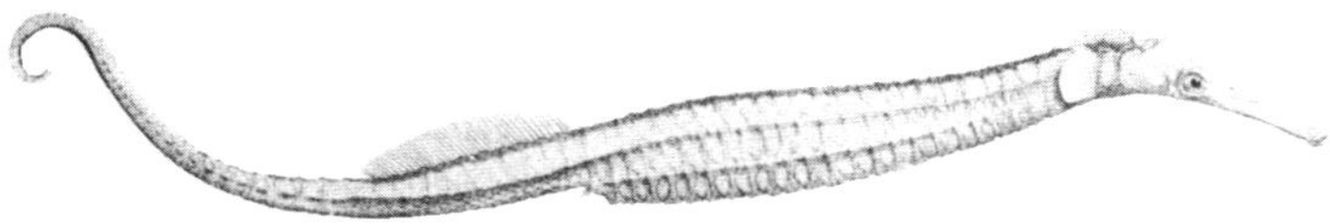

3

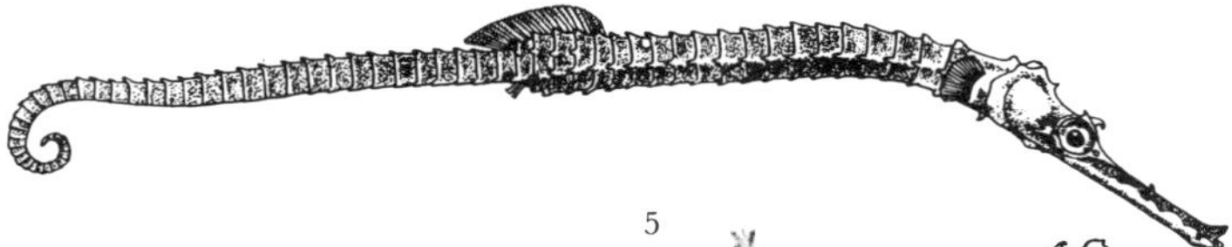

4

5

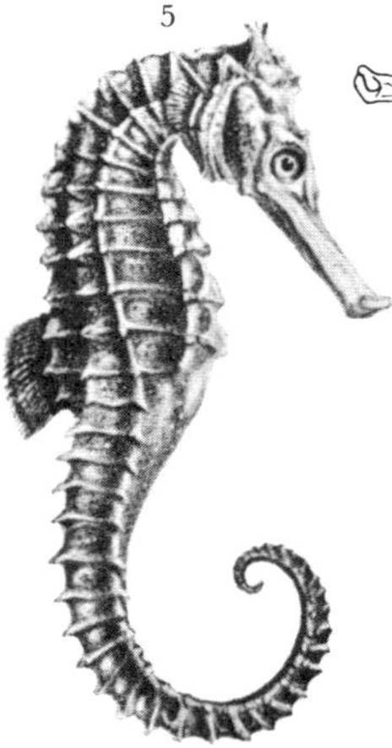

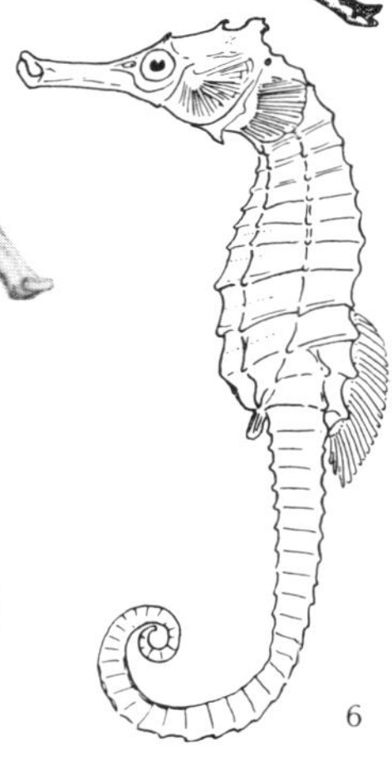

6

1. Wide-bodied Pipe-fish, *Stigmatophora nigra.* Top, lateral view of female; centre, ventral surface of female; bottom, ventral surface of male.
2. Horned Pipe-fish, *Syngnathoides biaculeatus.*
3. Red and Gold Sea-horse, *Solegnathus dunckeri,* 46cm.
4. Slender Tangle-fish, *Haliïchthys taeniophora.*
5. Common Sea-horse, *Hippocampus whitei.*
6. Low-crowned Sea-horse, *Hippocampus planifrons,* 10cm.

PLATE 101

1. Southern Ribbon-fish, *Trachipterus jacksonensis,* 1.90m.
2. Oar-fish, *Regalecus pacificus,* 4.15m.
3. Roseate Boarfish, *Antigonia rubicunda,* 6cm.
4. Knight-fish, *Cleidopus gloriamaris,* 18cm.
5. Nannygai, *Centroberyx affinis.*

PLATE 102

1

2

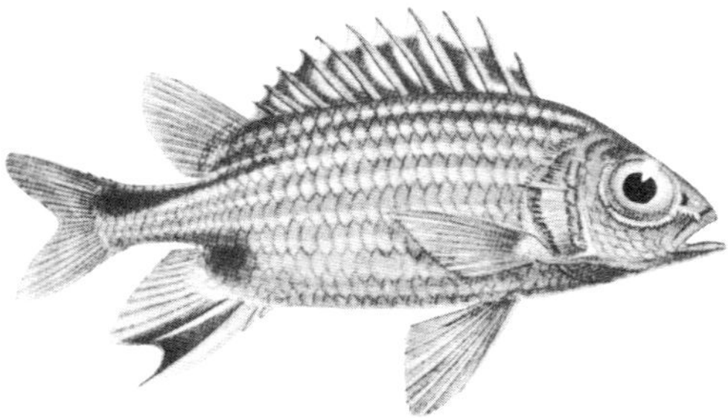

3

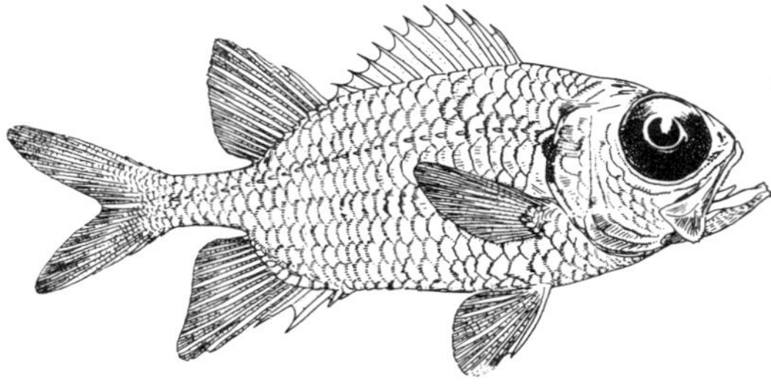

4

1. Roughy, *Trachichthys australis,* 19cm.
2. Slender Roughy, *Optivus elongatus.*
3. Red and White Soldier-fish, *Adioryx cornutum,* 13cm.
4. Blunt-nosed Soldier-fish, *Myripristis hexagonus,* 16cm.

PLATE 103

1

3

4

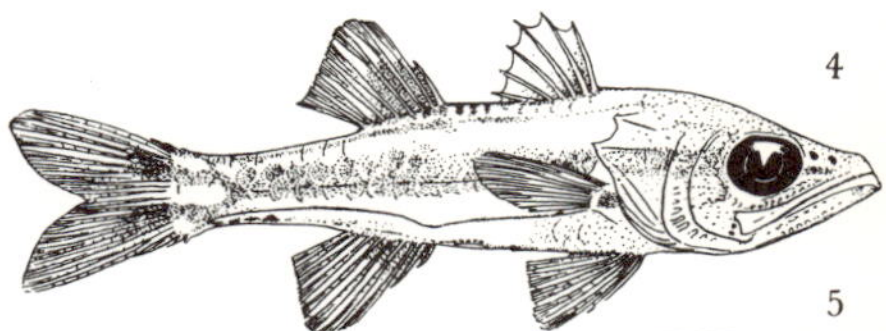

5

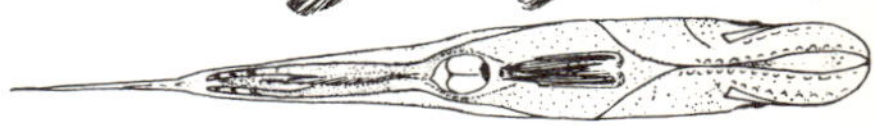

2

1. John Dory, *Zeus faber*.
2. Rose-bellied Siphon-fish, *Siphamia roseigaster*, 6cm.
3. Bar-cheeked Cardinal-fish, *Apogon bandanensis*, 11cm.
4. Siphon-fish, *Siphamia cuneiceps*, holotype, Queensland female, 5cm.
5. Ventral surface of same specimen.

PLATE 104

1

2

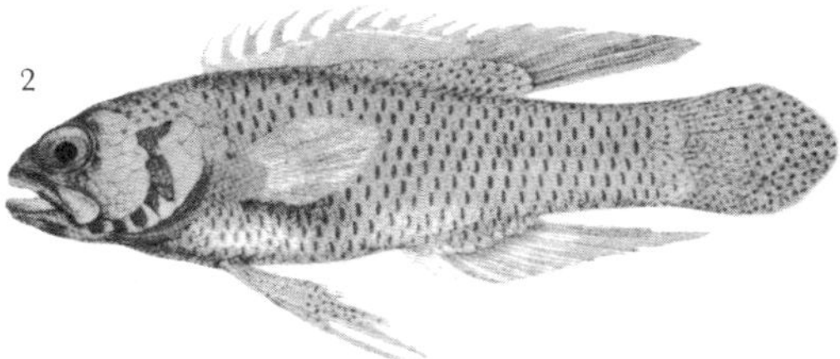

3

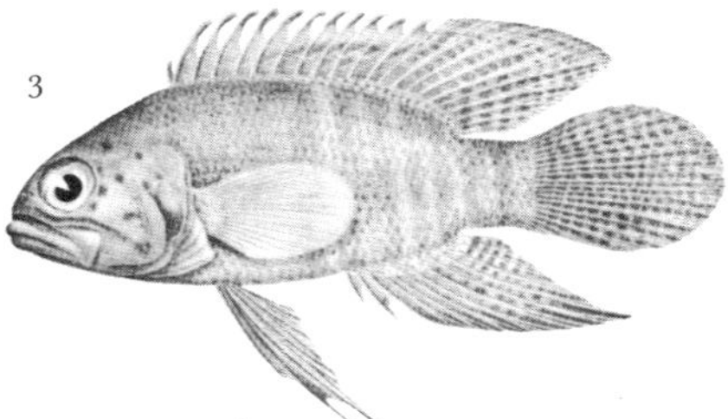

4

5

1. Five-striped Percelle, *Cheilodipterus quinquelineatus,* 9cm.
2. Blue-spotted Long-fin, *Plesiops coeruleolineatus.*
3. Blue-tip Long-fin, *Paraplesiops poweri,* 16cm.
4. Sand Bass, *Psammoperca waigiensis,* 30cm.
5. Giant Perch or Palmer, *Lates calcarifer,* about 38cm.

PLATE 105

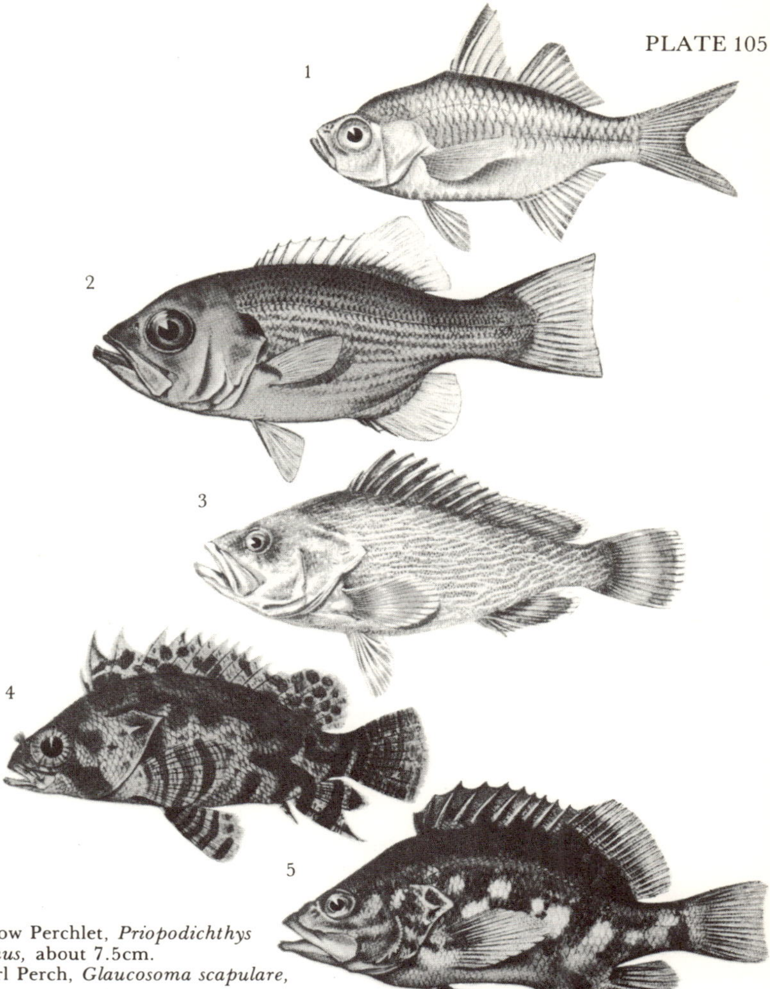

1. Yellow Perchlet, *Priopodichthys marianus,* about 7.5cm.
2. Pearl Perch, *Glaucosoma scapulare,* 24cm.
3. Maori Rock Cod, *Epinephelus undulatostriatus,* 29cm.
4. Pretty-fins, *Centrogenys vaigiensis,* 14cm.
5. Jameson's Rock Cod, *Ellerkeldia jamesoni,* 10cm.

PLATE 106

1

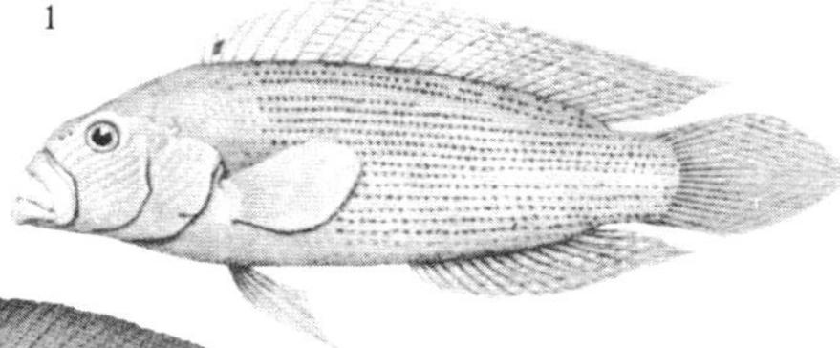

2

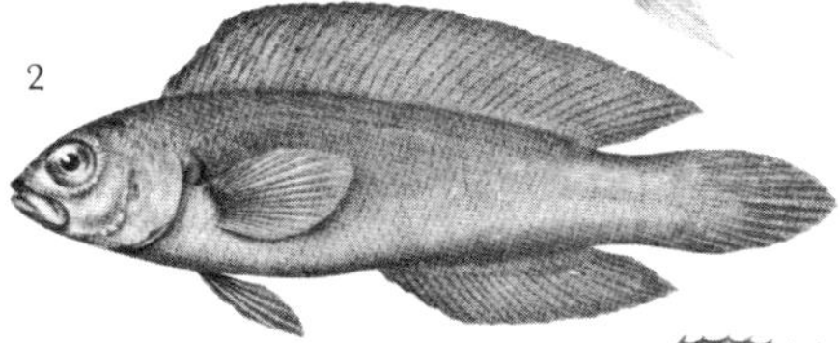

3

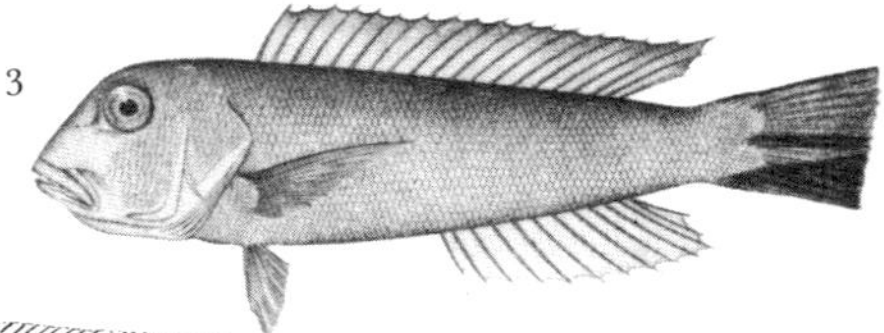

4

5

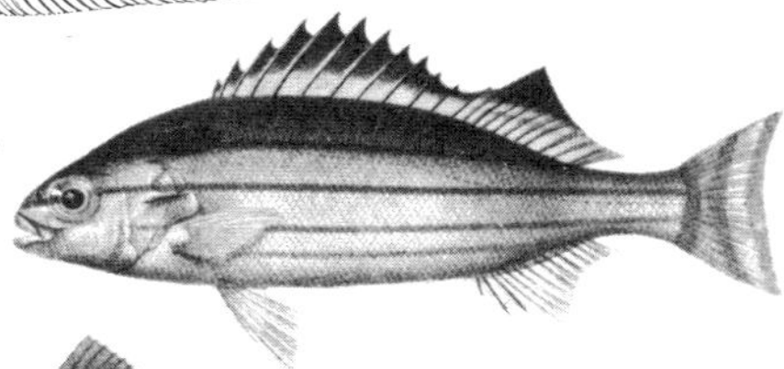

6

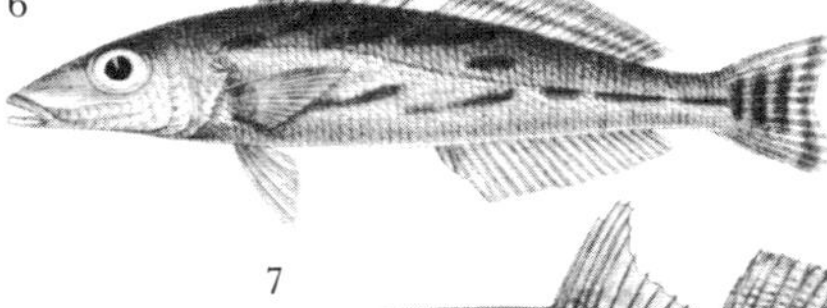

7

1. Short-finned Cichlops, *Dampieria lineata,* 25cm.
2. New Holland Cichlops, *Pseudochromis novaehollandiae,* 19cm.
3. *Branchiostegus ilocanus,* an allied species to *B. wardi.*
4. Band-fish, *Cepola australis.*
5. Six-lined Perch, *Pelates sexlineatus,* 19cm.
6. Trumpeter Whiting, *Sillago maculata.*
7. Sand Whiting, *Sillago ciliata,* 31cm.

PLATE 107

1

2

3

4

5

6

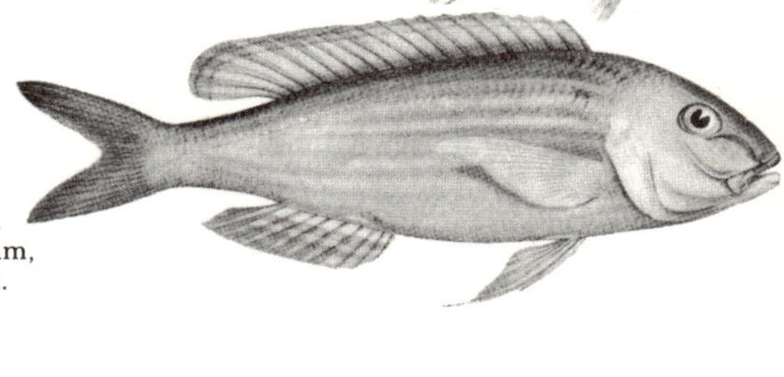

1. Sergeant-fish, *Rachycentron canadus,* juv., 25cm.
2. Sergeant-fish, adult, 1m.
3. Tailor, *Pomatomus saltatrix,* 38cm.
4. Blackfish, *Girella tricuspidata,* 29cm.
5. Topsail Drummer, *Kyphosus cinerascens,* 23cm.
6. Theodore's Butterfly Bream, *Nemipterus theodorei,* 25cm.

PLATE 108

1
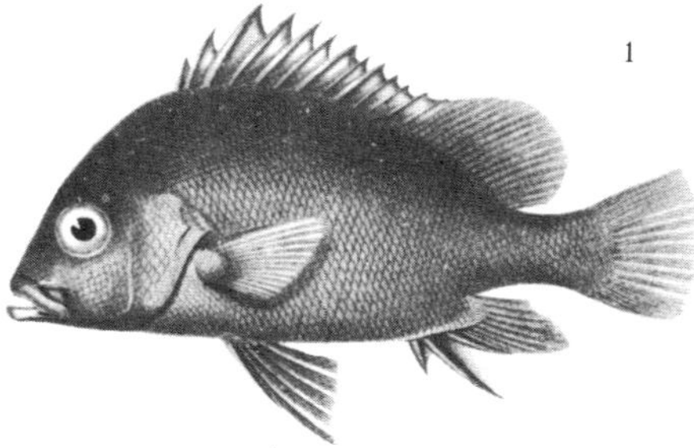

1. Brown Sweet-lips,
Plectorhynchus nigrus, 21cm.
2. Triple-tail,
Lobotes surinamensis, 33cm.
3. Cape Moreton Sea Bream,
Dentex spariformis, 29cm.
4. Cast of head of Snapper,
Chrysophrys auratus.
See also PLATE 27.
5. Silver Teraglin,
Otolithes ruber, 25cm.
6. Teraglin, *Altractoscion aequidens.*

2

3
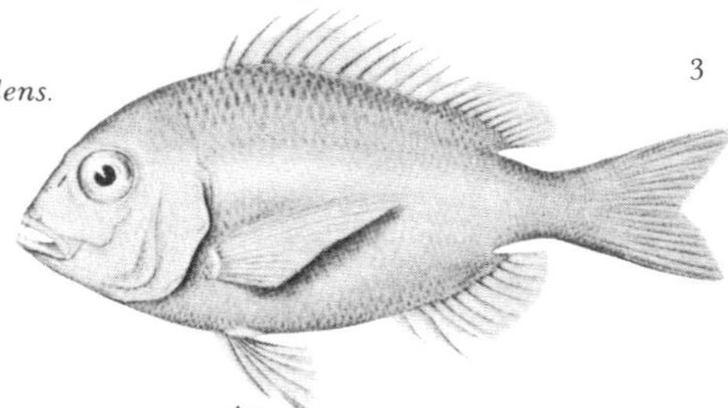

5

4

6

PLATE 109

1

2

3

4

5

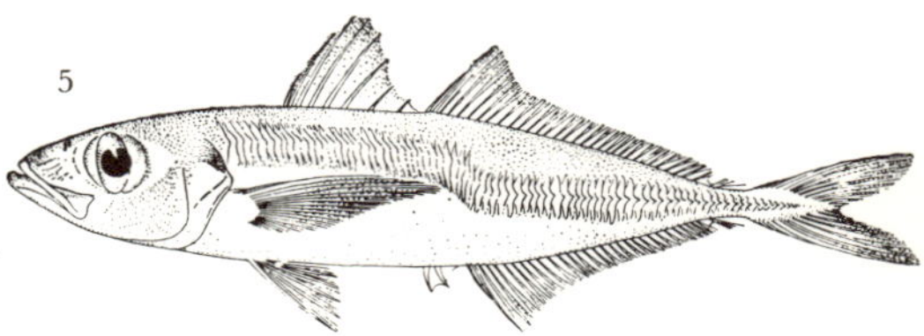

1. Jewfish or Mulloway, *Argyrosomus hololepidotus*, juv., 28cm.
2. Little Jewfish, *Johneiops vogleri*, 25cm.
3. Silver Jewfish, *Nibea soldado*, 35cm.
4. Northern Mackerel Scad, *Decapterus russellii*, 26cm.
5. Scad, *Trachurus mccullochi*, 21cm.

PLATE 110

1. White Trevally,
Caranx nobilis, 25cm.
2. Cale-cale Trevally,
Ulua mandibularis, 30cm.
3. Diamond Trevally,
Alectis indica, 36cm.
4. Swallow-tail or Dart,
Trachinotus russelli, 26cm.
5. Pilot-fish,
Naucrates ductor, 2.5cm.
6. Pilot-fish, 4cm.

PLATE 111

1

2

4

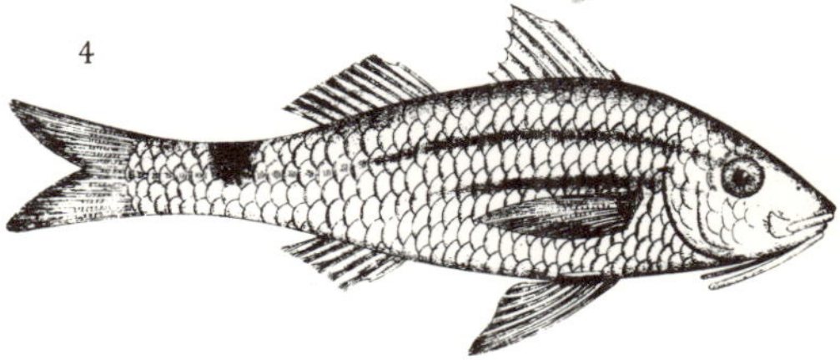

6

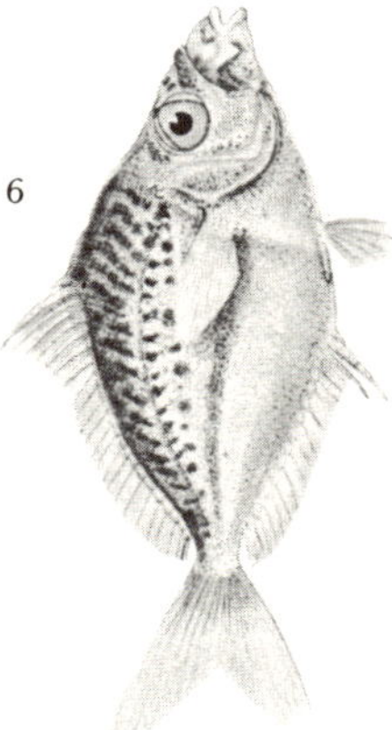

5

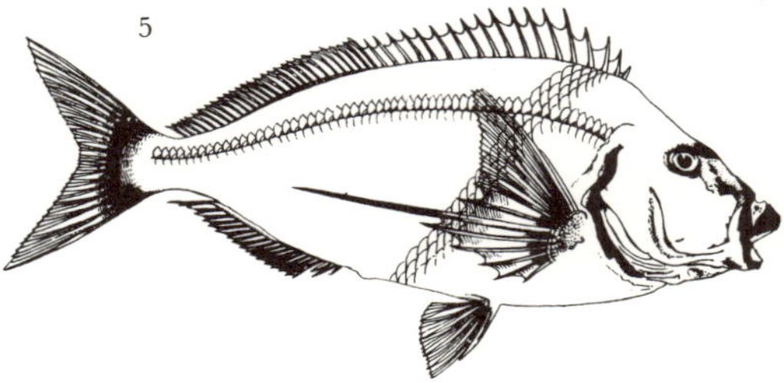

3

1. Mottled Goat-fish, *Upeneus tragula*, 21cm.
2. Gold-striped Goat-fish, *Mulloidichthys vanicolensis.*
3. Magpie Morwong, *Cheilodactylus gibbosus.*
4. Black-spot Goat-fish, *Parupeneus signatus,* 16cm.
5. Morwong, *Nemadactylus douglasii.*
6. Black-banded Pony-fish, *Equulites moretoniensis,* 7.5cm.

PLATE 112

2

1

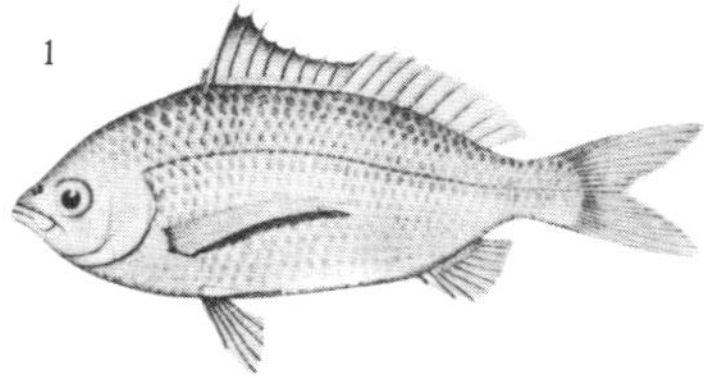

3

5

6

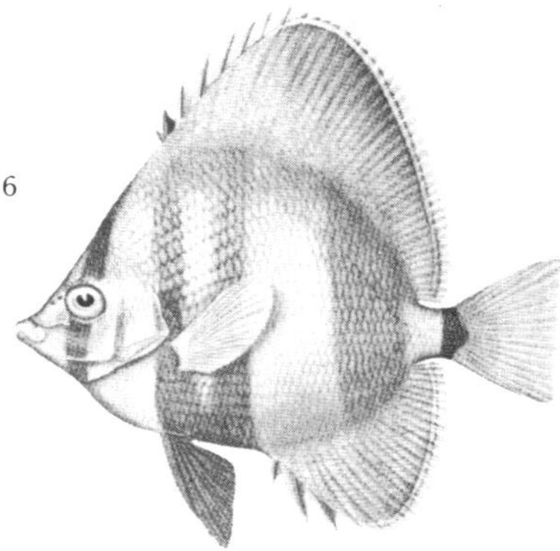

7

4

1. Silver-belly, *Gerres filamentosus,* 16cm.
2. Thread-fin Silver-belly, *Gerres punctatus,* 21cm.
3. Southern Butter-fish, *Selenotoca multifasciata,* 38cm.
4. Stripey, *Microcanthus strigatus,* 10.5cm.
5. Truncate Coral-fish, *Chelmonops truncatus,* 20cm.
6. High-fin Coral-fish, *Coradion altivelis,* 13.5cm.
7. Rainford's Butterfly-fish, *Chaetodon rainfordi,* 11cm.

PLATE 113

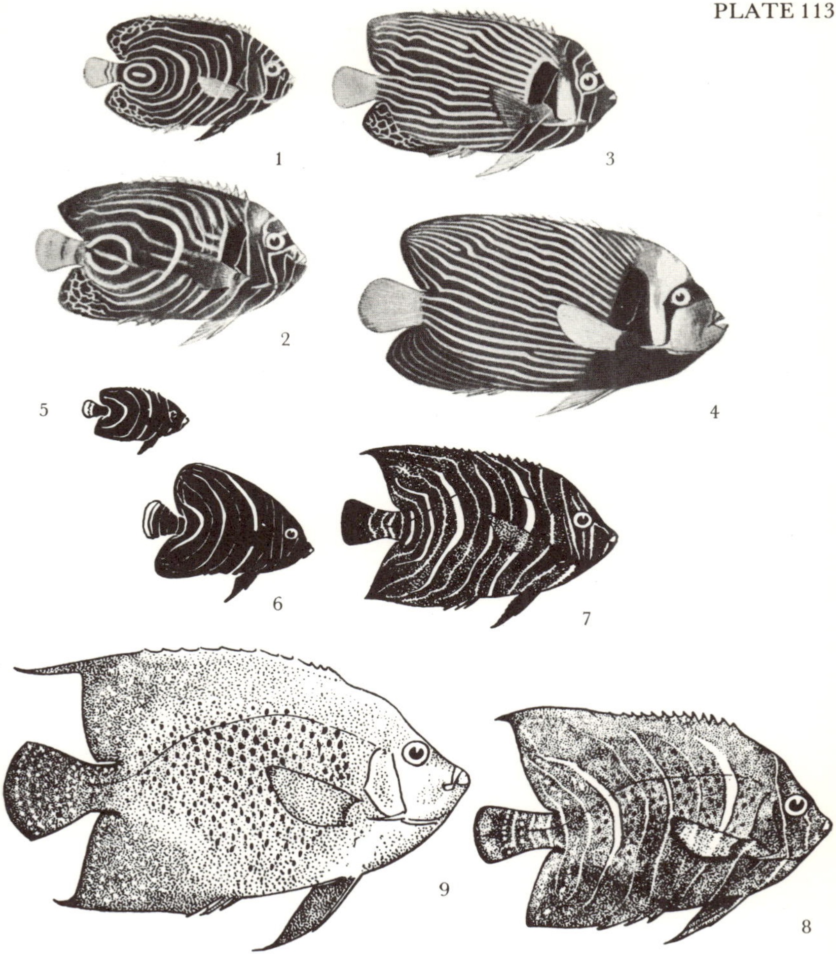

Emperor Angel-fish, *Pomacanthus imperator,* showing pattern development from juvenile to adult (1 to 4).
Zebra Angel-fish, *Pomacanthus semicirculatus,* showing pattern development (5 to 9).

PLATE 114

1

2

3

4

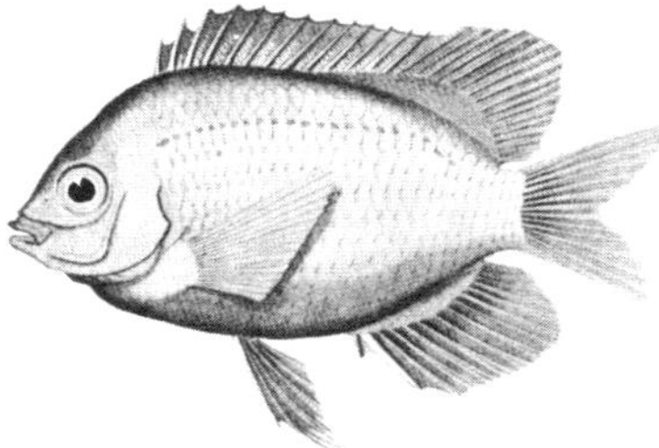

6

7

5

1. Six-banded Angel-fish, *Euxiphipops sexstriatus,* 25cm.
2. Black and Gold Angel-fish, *Centropyge bicolor.*
3. Old Wife, *Enoplosus armatus.*
4. Spine-cheeked Anemone-fish, *Premnas biaculeatus,* 11cm.
5. Spotty-tail, *Acanthochromis polyacanthus,* 13cm.
6. Green Puller, *Mecaenichthys immaculatus,* 14cm.
7. Banded Humbug, *Dascyllus aruanus,* 7.5cm.

PLATE 115

2

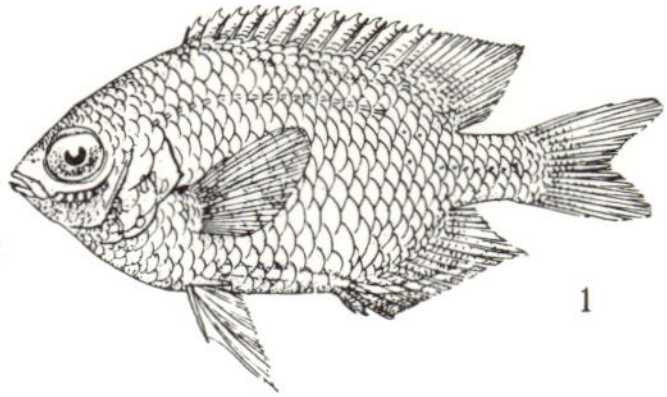

1

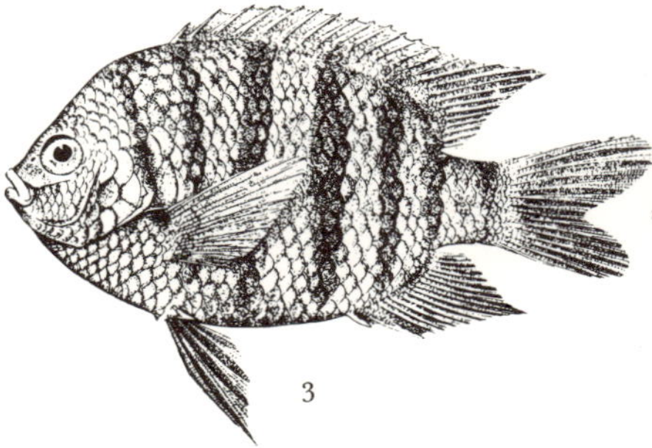

3

4

5

7

1. Yellow Damsel-fish, *Pomacentrus moluccensis,* 3cm.
2. Orange-tipped Damsel-fish, *Stegastes apicalis,* 11 cm.
3. Palmer's Damsel-fish, *Abudefduf bengalensis,* 15cm.
4. Clouded Damsel-fish, *Amblyglyphidodon curacao,* 10cm.
5. Orange-tailed Damsel-fish, *Neopomacentrus nemurus,* 7.5cm.
6. Big-lip Damsel-fish, *Cheiloprion labiatus,* 9cm.
7. Large-scale Parma, *Parma oligolepis,* 18cm.

6

PLATE 116

1

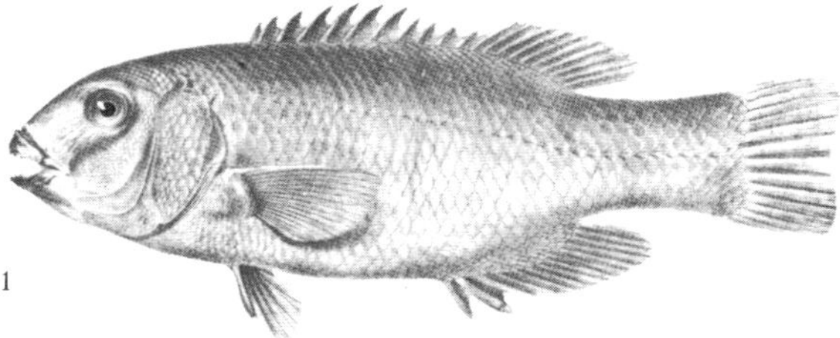

2

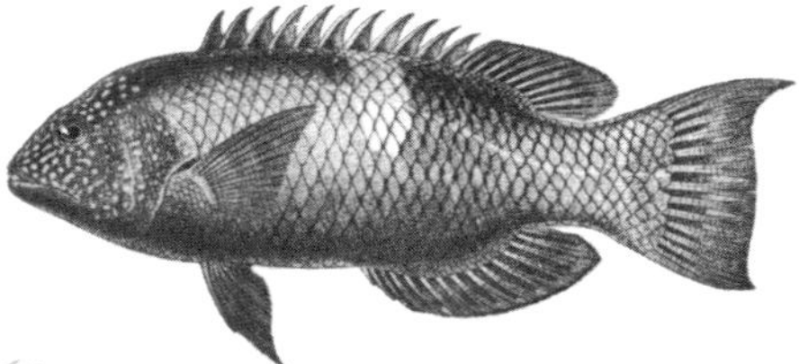

3

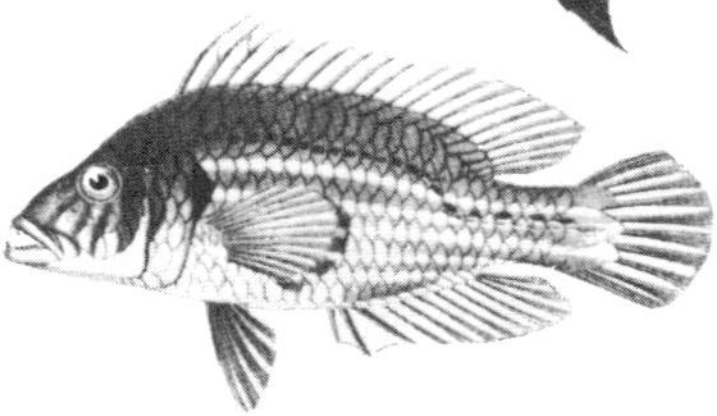

4

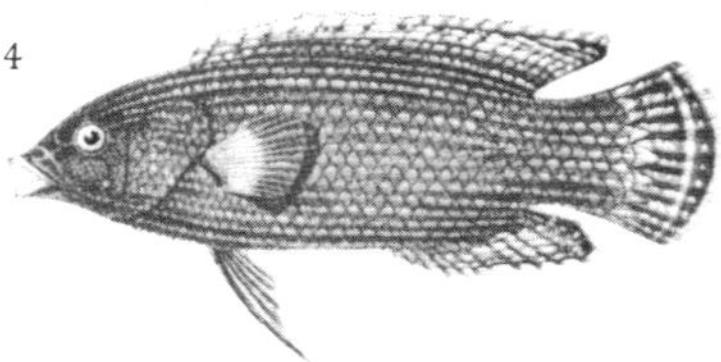

5

7

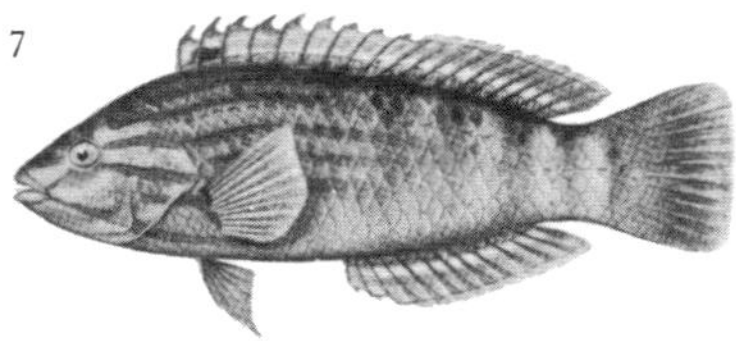

6

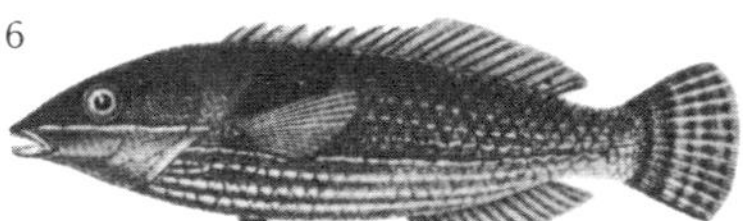

1. Giant Pig-fish, *Achoerodus unicolor,* 1.10m.
2. Golden-spot Pig-fish, *Bodianus perditio,* 33cm.
3. Cockatiel-fish, *Pteragogus flagellifera,* 14cm.
4. Tube-lip, *Labrichthys unilineata,* 13cm.
5. Blue Streak, *Labroides dimidiatus,* 8cm.
6. Lined Rainbow-fish, *Stethojulis strigiventer,* 10cm.
7. Günther's Rainbow-fish, *Pseudolabrus guentheri.*

PLATE 117

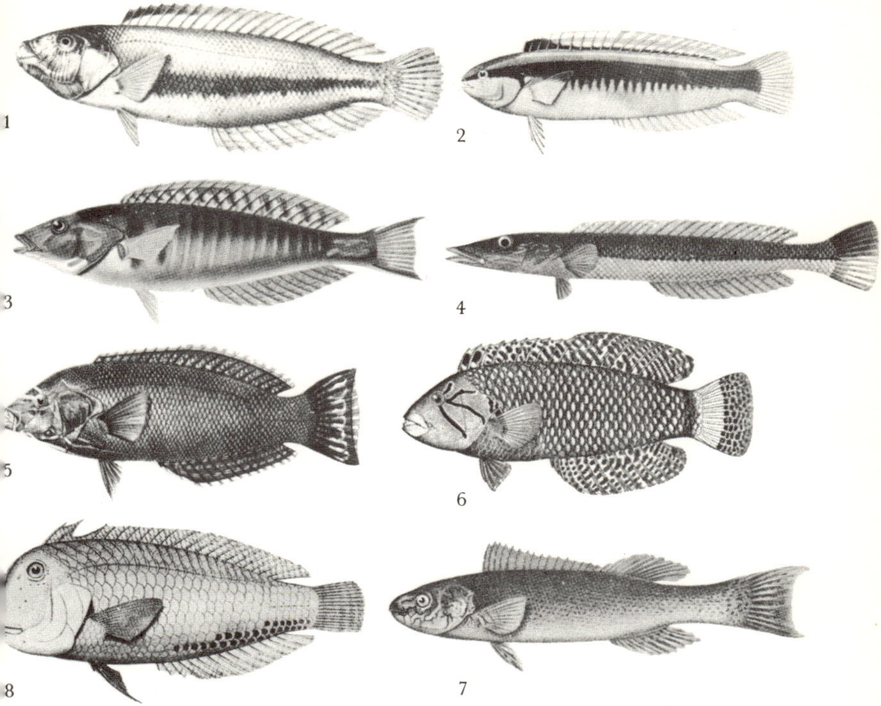

1. Maori, *Ophthalmolepis lineolatus,* 40cm.
2. Comb-fish, *Coris picta,* 22cm.
3. Ringed Wrasse, *Hologymnosus semidiscus.*
4. Sharp-nosed Wrasse, *Cheilio inermis,* 25cm.
5. Scribbled Wrasse, *Anampses geographicus,* 20cm.
6. Bar-cheeked Wrasse, *Hemipteronotus taeniourus,* 25cm.
7. Keel Head, *Hemipteronotus jacksonensis*, 18.5cm.
8. Herring Cale, *Olisthops cyanomelas,* 29cm.

PLATE 118

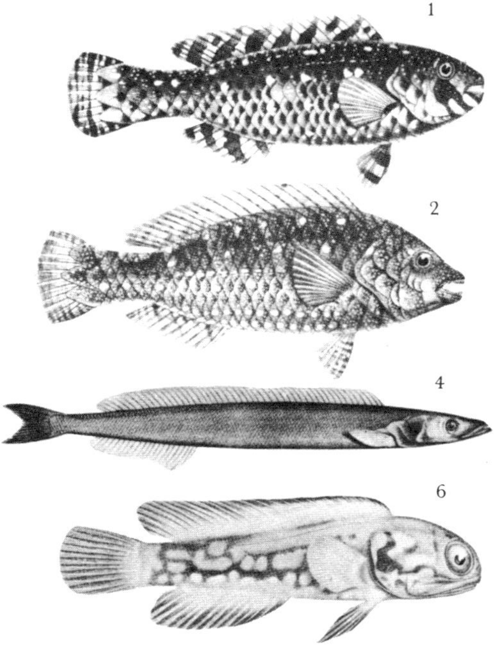

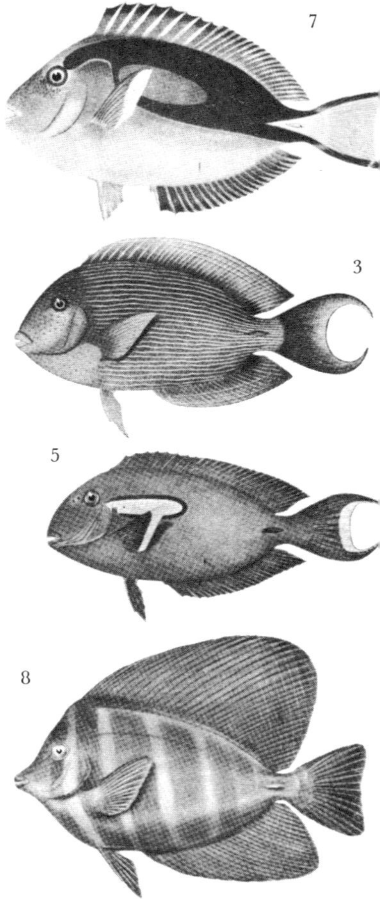

1. Marbled Parrot-fish, *Leptoscarus vaigiensis,* 23cm.
2. Half-toothed Parrot-fish, *Calotomus spinidens,* 15cm.
3. Bristle-toothed Surgeon-fish, *Ctenochaetus strigosus.*
4. Sand Lance, *Bleekeria vaga,* 16cm.
5. Orange-spot Surgeon-fish, *Acanthurus olivaceus,* about 23cm.
6. Harlequin Smiler, *Merogymnus eximius,* 30cm.
7. Flag-tail Surgeon-fish, *Paracanthurus hepatus,* 15cm.
8. Sail-finned Surgeon-fish, *Zebrasoma veliferum,* 30cm.

PLATE 119

1. Fox-face, *Siganus vulpinus.*
2. Moorish Idol, *Zanclus cornutus,* 6cm.
3. Black Spinefoot, *Siganus rivulatus,* 21cm.
4. Cutlass-fish, *Lepturacanthus savala,* 30cm.
5. Black Marlin, *Makaira indica,* 4.25m.

PLATE 120

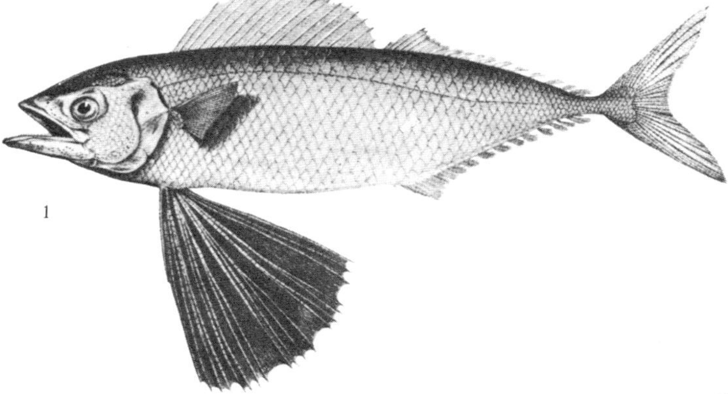

1

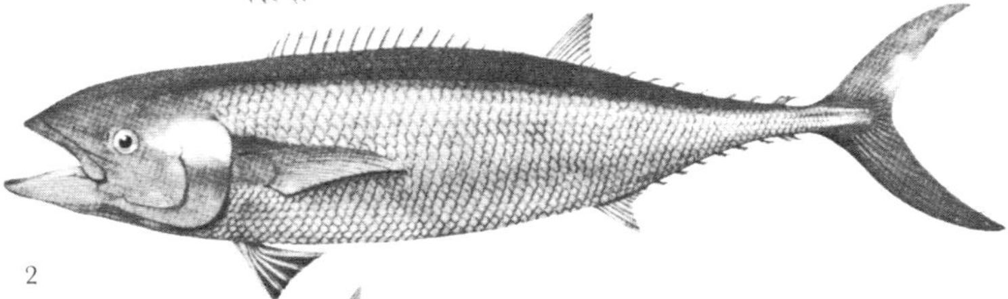

2

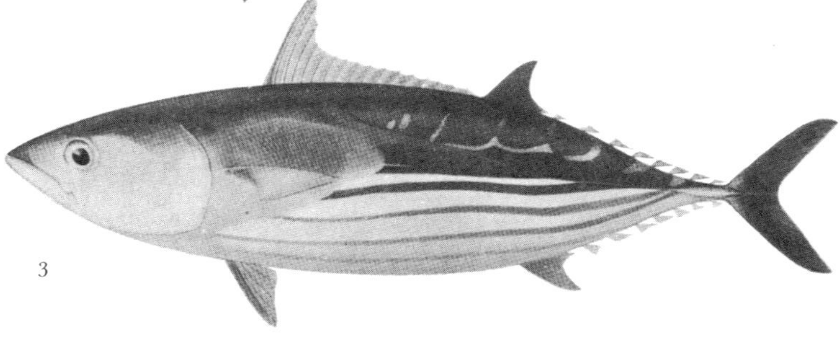

3

1. Big-scaled Mackerel, *Gasterochisma melampus,* juv., 20cm.
2. Big-scaled Mackerel, adult, 1.75m.
3. Skipjack Tuna, *Katsuwonus pelamis,* 80cm.

PLATE 121

1

2

3

4

1. Narrow-barred Mackerel, *Scomberomorus commerson,* juv., 40cm.
2. Narrow-barred Mackerel, adult, 1.05m.
3. Queensland School Mackerel, *Scomberomorus queenslandicus,* juv., 20cm.
4. Queensland School Mackerel, adult, 60cm.

PLATE 122

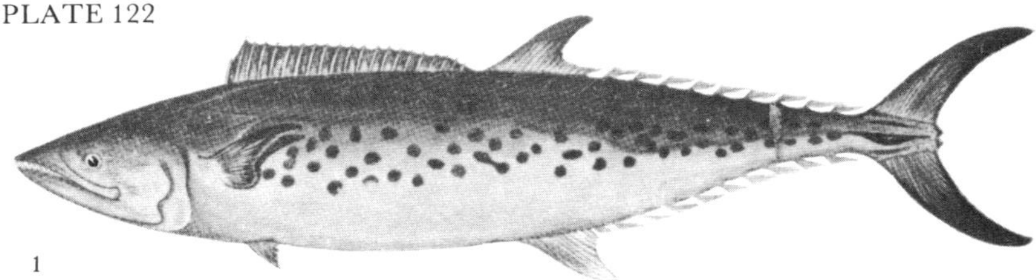

1

2

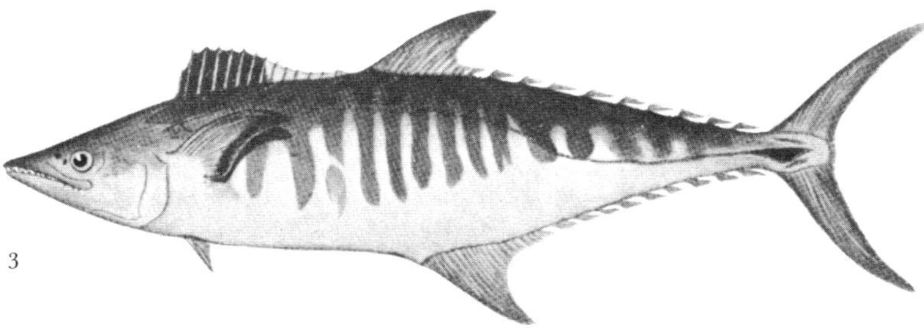

3

4

1. Spotted Mackerel, *Scomberomorus niphonius,* 83cm.
2. Broad-barred Mackerel, *Scomberomorus semifasciatus,* juv., 16cm.
3. Broad-barred Mackerel, adult, 50cm.
4. Wahoo or Peto, *Acanthocybium solandri,* 1.10m.

PLATE 123

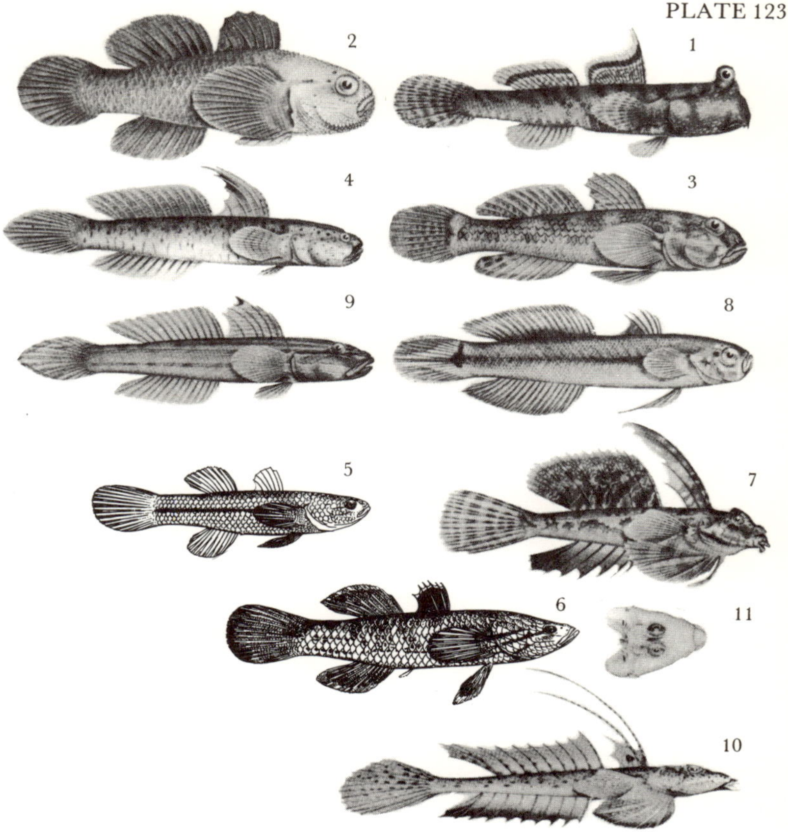

1. Mud-hopper, *Periophthalmus regius,* 9cm.
2. Redhead, *Paragobiodon echinocephalus,* 2.5cm.
3. Oyster Goby, *Arenigobius leftwichi,* 6cm.
4. Crested Mud Goby, *Cryptocentroides cristatus,* 9cm.
5. Mud Gudgeon, *Ophiocara aporos,* juv.
6. Mud Gudgeon, adult.
7. Fingered Dragonet, *Dactylopus dactylopus,* female, 11cm. (See also PLATE 53.)
8. Cobra-fish, *Parioglossus rainfordi,* 4cm.
9. Lined Gudgeon, *Valenciennea muralis,* 11cm.
10. Rough-headed Dragonet, *Callionymus limiceps,* male, 19.5cm.
11. Head of same specimen.

PLATE 124

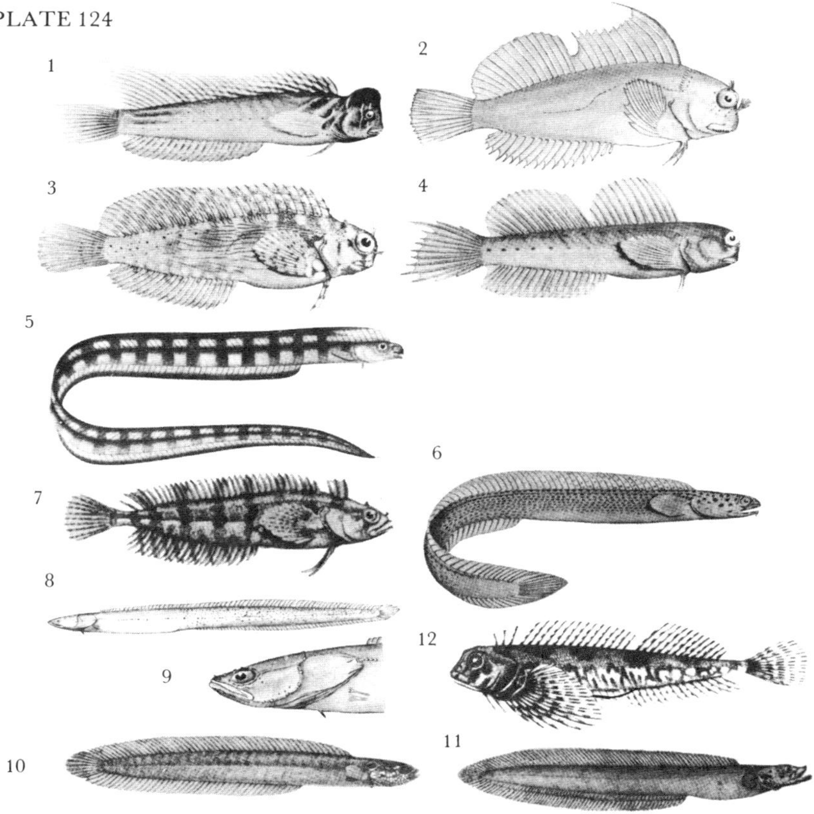

1. Oyster Blenny, *Omobranchus anolius,* 7cm.
2. Collared Blenny, *Cirripectes variolosus,* 7cm.
3. Lineated Blenny, *Salarias fasciatus,* 7cm.
4. Many-toothed Blenny, *Ecsenius mandibularis,* female, 5cm.
5. Hair-tailed Blenny, *Xiphasia setifer,* 50cm.
6. Spotted Eel Blenny, *Notograptus guttatus,* 11cm.
7. Long-nosed Weed-fish, *Heteroclinus nasutus,* 7cm.
8. Snake Blenny, *Peronedys anguillaris,* 7.5cm.
9. Enlarged lateral view of head (Snake Blenny).
10. Eel Blenny, *Blennodesmus scapularis,* 7.5cm.
11. Mud Blenny, *Congrogadus subducens,* 28cm.
12. Jumping Joey, *Lepidoblennius haplodactylus,* 10cm.

PLATE 125

1. Hump-back Messmate, *Carapus houlti,* 8cm.
2. Portuguese Man-O'-War Fish, *Nomeus albula,* 8cm.
3. Whitelegge's Blubber-fish, *Psenes whiteleggii,* 5cm.
4. Tallegalane or Sand Mullet, *Myxus elongatus.*
5. Freshwater Mullet, *Trachystoma petardi,* 35cm.

PLATE 126

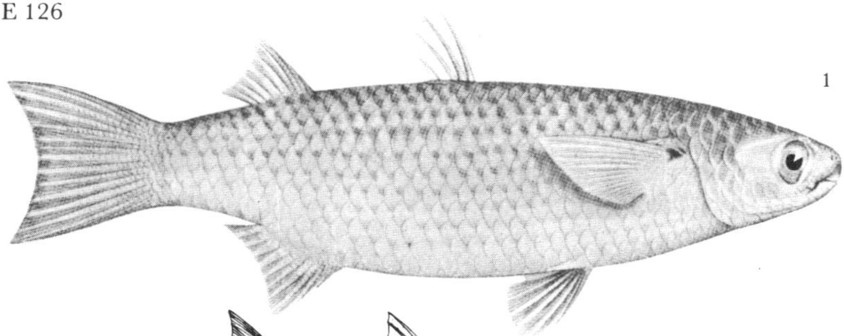

1

2

3

1. Fantail Mullet, *Mugil georgii*, 23cm.
2. Long-finned Mullet, *Liza strongylocephalus*, 23cm.
3. Flat-tail Mullet, *Liza dussumieri*, 26cm.

PLATE 127

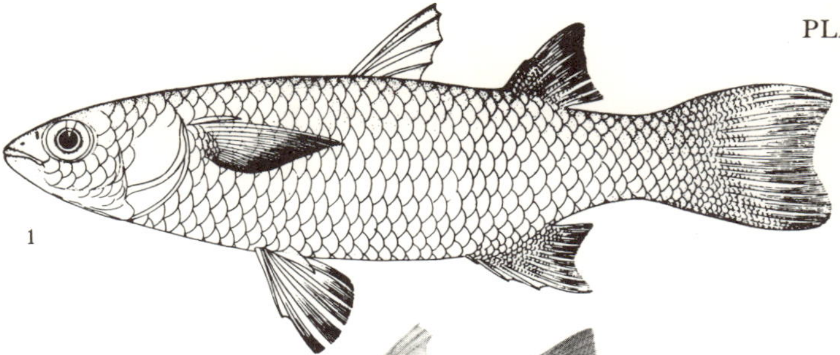

1

2

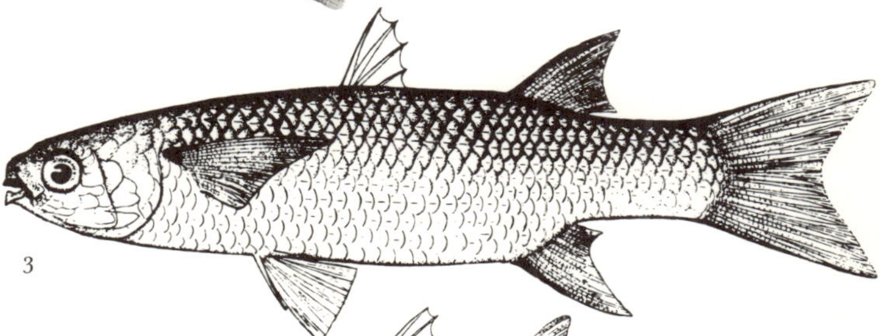

3

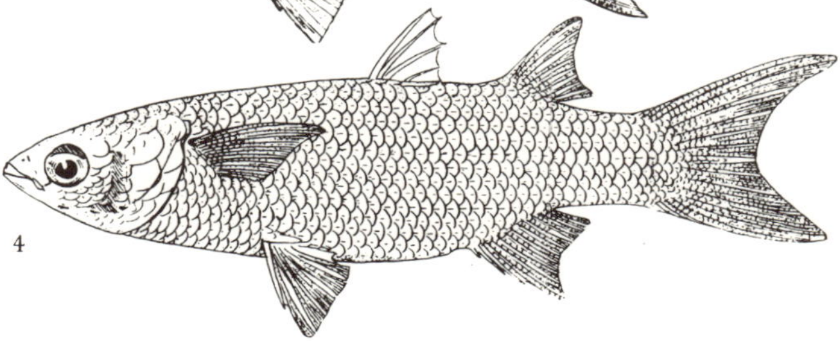

4

1. Tade Mullet, *Liza tade,* 8.5cm.
2. Sea Mullet, *Mugil cephalus.*
3. Warty-nosed Mullet, *Crenimugil crenilabis,* 20cm.
4. Ramsay's Mullet, *Liza ramsayi,* 20cm.

PLATE 128

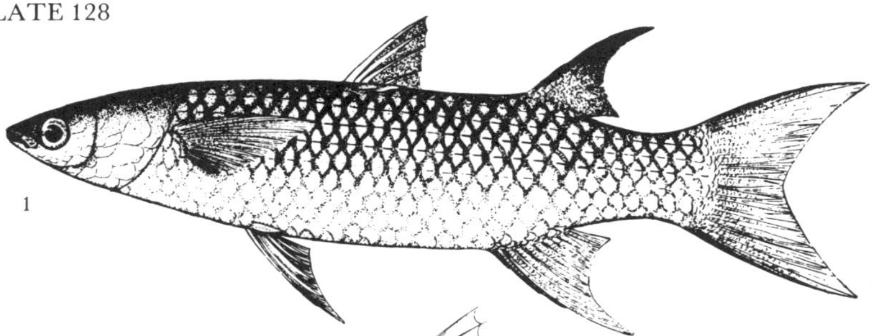

1

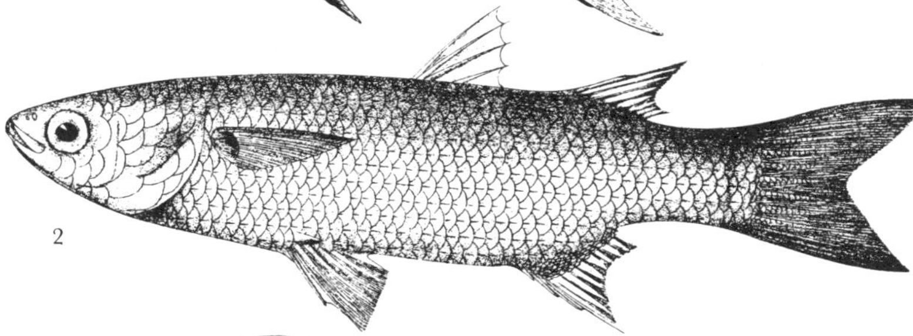

2

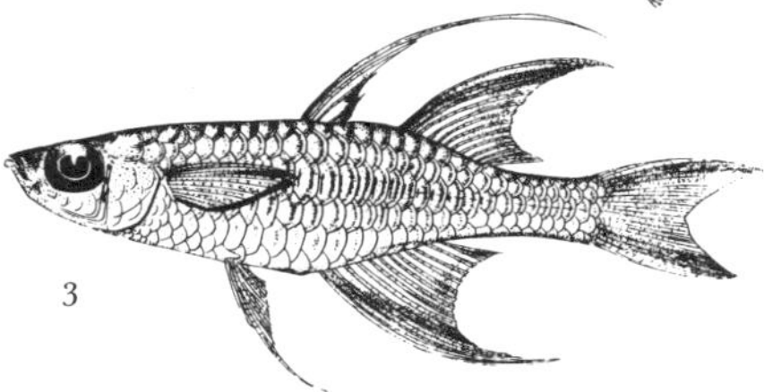

3

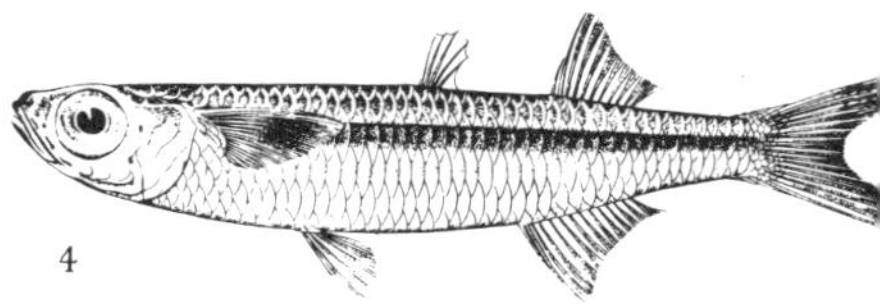

4

1. Basket Mullet, *Liza diadema,* 45cm.
2. Tiger Mullet, *Liza argentea,* 26cm.
3. Marine Blue-eye, *Pseudomugil signifer,* male, much enlarged.
4. Common Hardyhead, *Pranesus ogilbyi,* 17cm.

PLATE 129

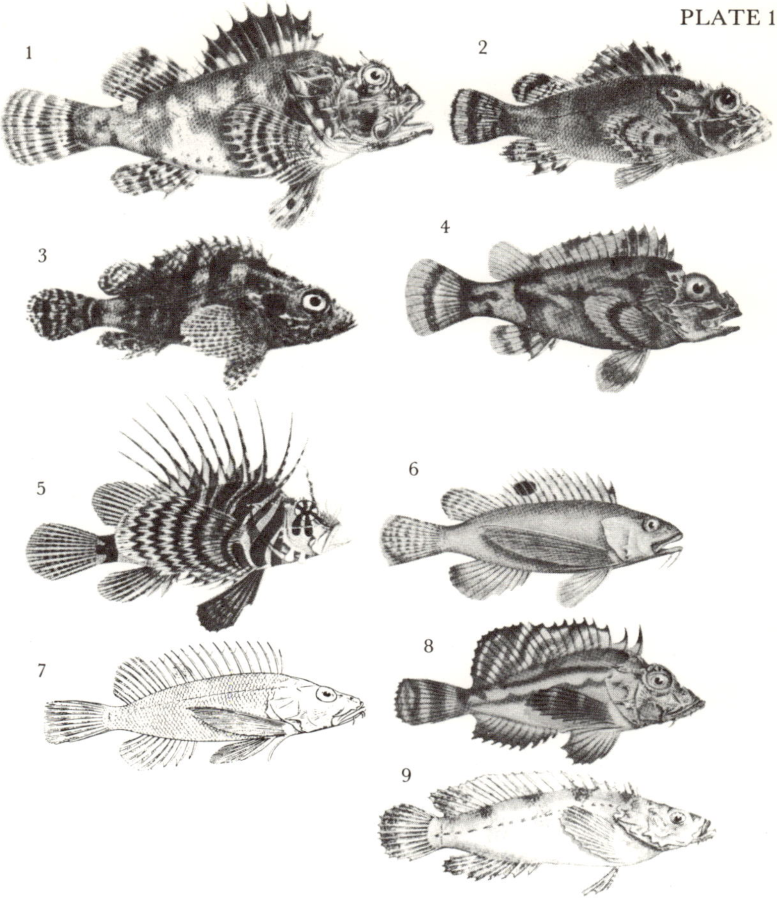

1. Red Scorpion Cod, *Scorpaena cardinalis*, 45cm.
2. Little Scorpion Cod, *Scorpaena bandanensis*, 11cm.
3. Guam Scorpion Cod, *Scorpaenodes guamensis*.
4. Marbled Fortesque, *Centropogon marmoratus*, 10cm.
5. Zebra Fire-fish, *Dendrochirus zebra*, 15cm.
6. Ocellated Wasp-fish, *Hypodytes carinatus*, 16cm.
7. Short-spined Wasp-fish, *Apistops caloundra*, 16cm.
8. Black-banded Wasp-fish, *Minous versicolor*, 10cm.
9. Brown Carpet-fish, *Peristrominous dolosus*, 8cm.

PLATE 130

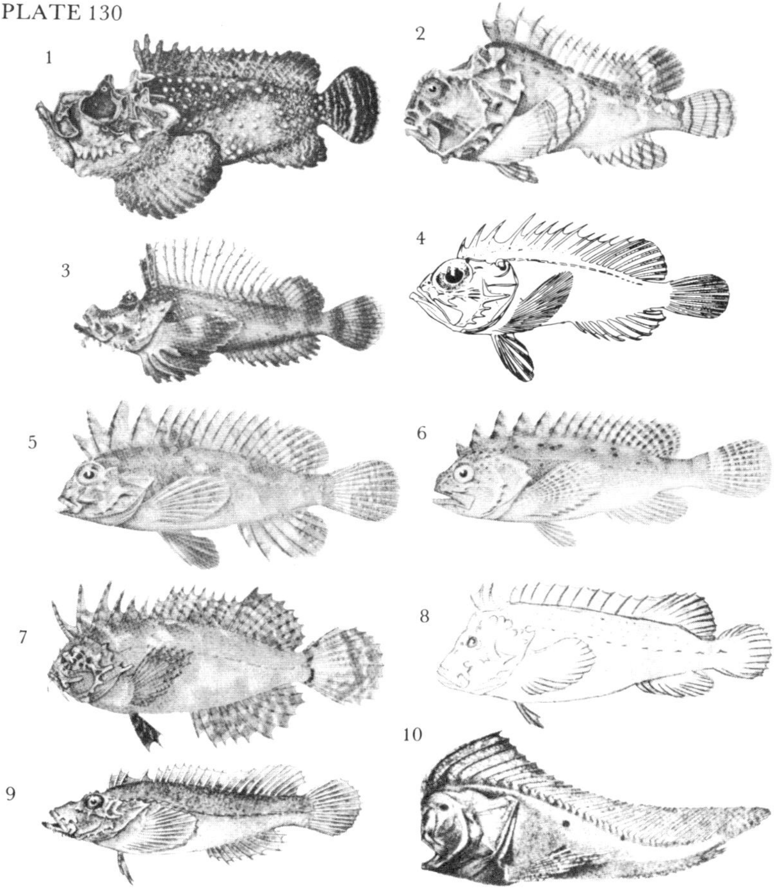

1. Stone-fish, *Synanceia horrida,* 23cm.
2. Monkey-fish, *Erosa erosa,* 7.5cm.
3. Bearded Ghoul, *Inimicus barbatus,* 16cm.
4. Rogue-fish, *Tetraroge leucogaster,* 6cm.
5. Freckled Wasp-fish, *Paracentropogon vespa,* 9cm.
6. Yellow Wasp-fish, *Liocranium scorpio,* 7.5cm.
7. Port Curtis Mossback, *Bathyaploactis curtisensis,* 6cm.
8. Northern Mossback, *Paraploactis trachyderma,* 15cm.
9. Sandpaper-fish, *Adventor elongatus,* 10cm.
10. Red Indian Fish, *Pataecus fronto,* nearly 23cm.

PLATE 131

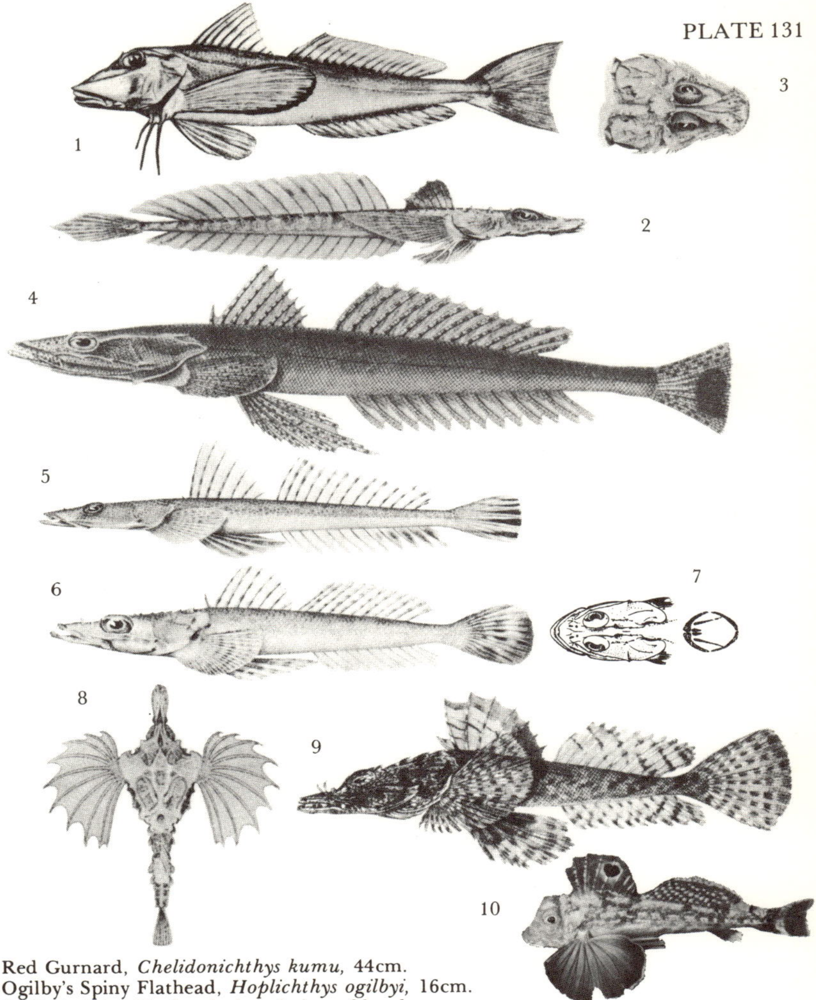

1. Red Gurnard, *Chelidonichthys kumu*, 44cm.
2. Ogilby's Spiny Flathead, *Hoplichthys ogilbyi*, 16cm.
3. Ogilby's Spiny Flathead, dorsal view of head.
4. Dusky Flathead, *Platycephalus fuscus*.
5. Sand Flathead, *Platycephalus arenarius*, 26cm.
6. Harris's Flathead, *Platycephalus harrisii*, about 18cm.
7. Harris's Flathead, dorsal view of head.
8. Short-tailed Dragon-fish, *Pegasus volitans*, 7.5cm.
9. Rock Flathead, *Platycephalus cirronasus*, 23.5cm.
10. Butterfly Gurnard, *Poratrigla papilio*, 18cm.

PLATE 132

2

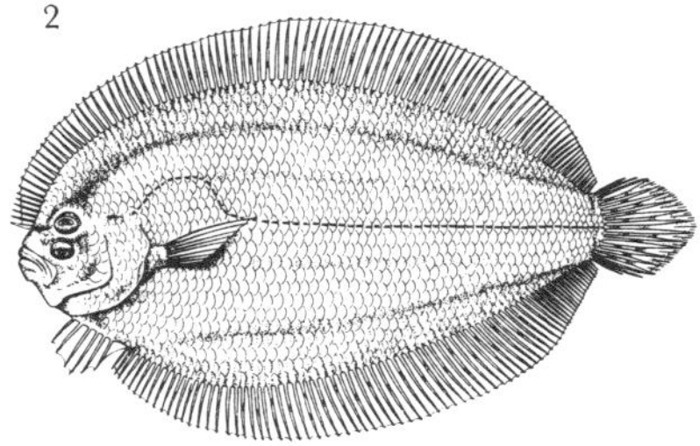

1

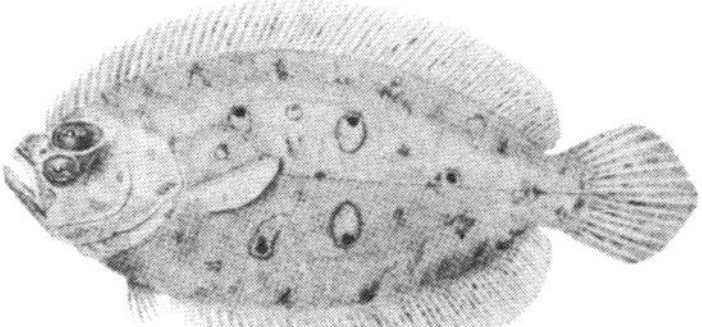

3

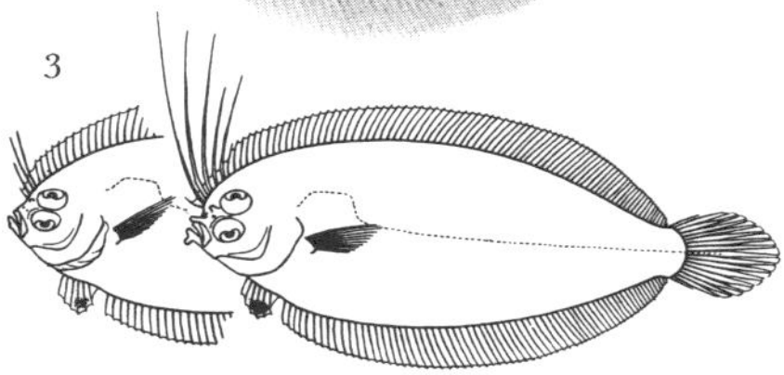

5

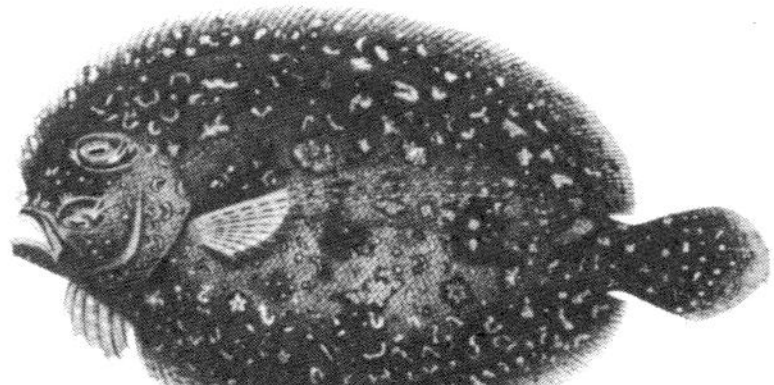

6

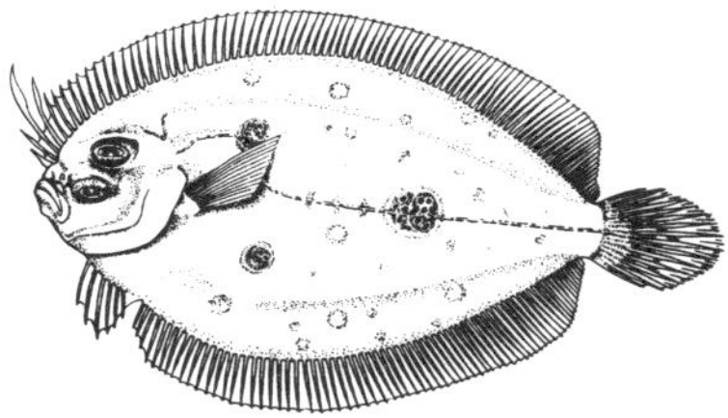

4

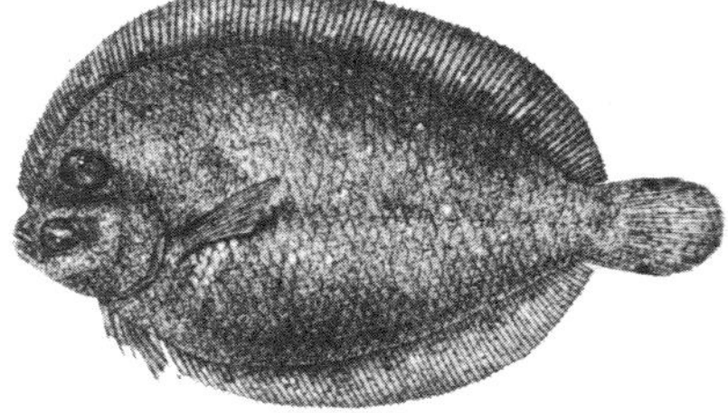

7

1. Small-toothed Flounder, *Pseudorhombus jenynsii,* 14cm.
2. Fison's Flounder, *Arnoglossus fisoni,* 13cm.
3. Crested Flounder, *Lophonectes gallus,* 15cm, showing male (right) and head of female (left).
4. Spotted-tail Flounder, *Engyprosopon grandisquama,* 10cm.
5. Panther Flounder, *Bothus pantherinus,* 19cm.
6. Feather-fin Flounder, *Grammatobothus pennatus,* 18.5cm.
7. Cockatoo Flounder, *Samaris cacatuae,* 17cm.

PLATE 133

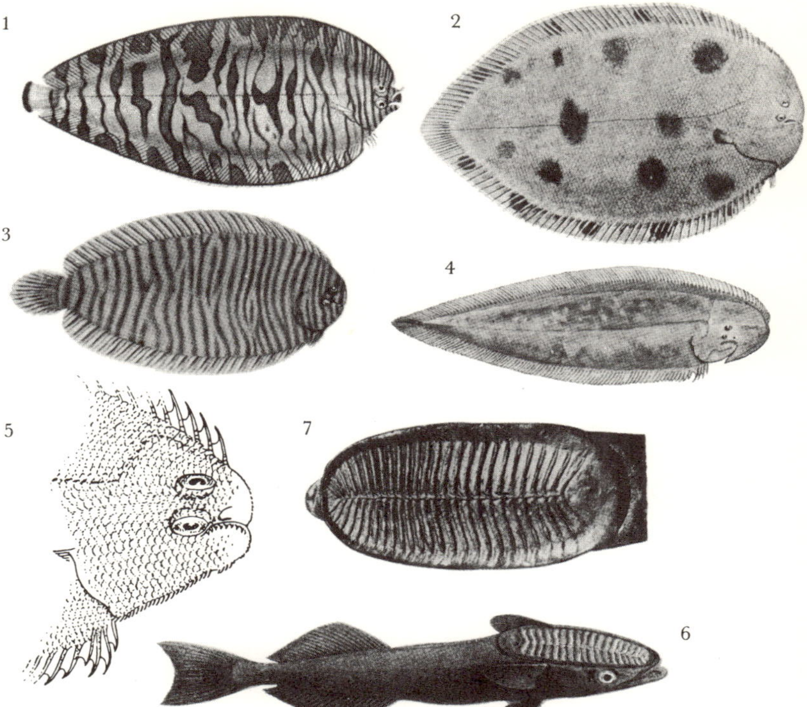

1. Tiger Sole, *Aesopia heterorhinos,* 14cm.
2. Black Sole, *Achylopa nigra,* 28cm.
3. Narrow-banded Sole, *Aseraggodes macleayanus,* 18cm.
4. Lemon Tongue Sole, *Paraplagusia unicolor,* 13cm.
5. Head of Sharp-headed Sole, *Phyllichthys sclerolepis.* (See also PLATE 63.)
6. Short Sucking-fish, *Remora remora.*
7. Head of another species of *Remora,* showing "sucker".

PLATE 134

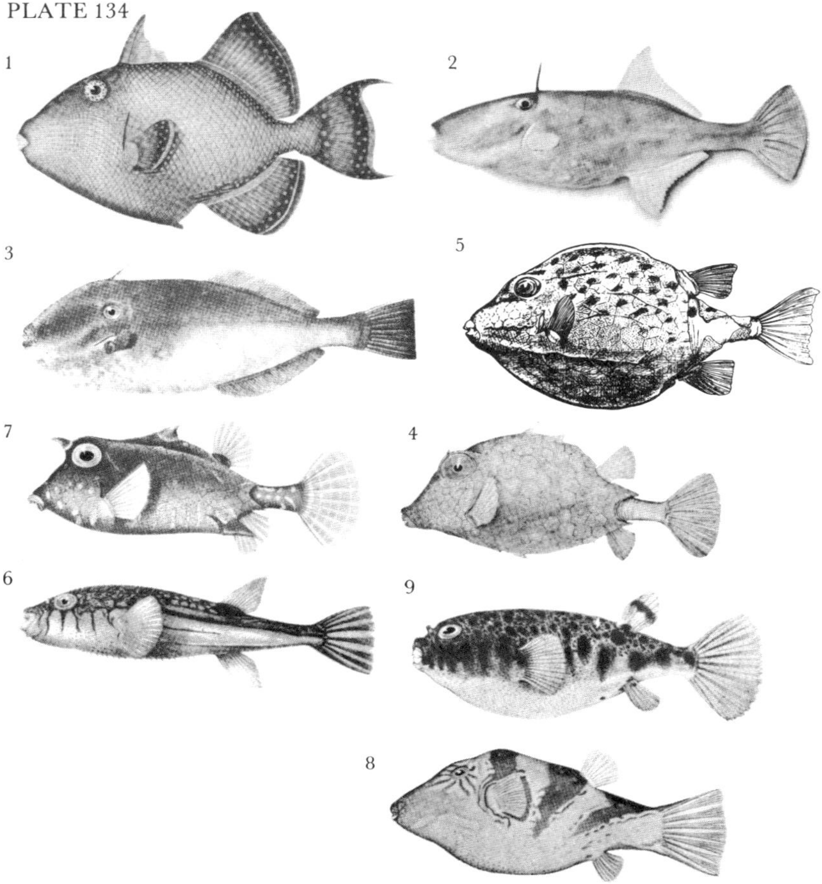

1. Yellow-spotted Trigger-fish, *Pseudobalistes fuscus.*
2. Chinaman Leather-jacket, *Nelusetta ayraudi.*
3. Unicorn Leather-jacket, *Alutera monoceros.*
4. Turret-fish, *Rhinesomus reipublicae,* about 10cm.
5. Robust Box-fish, *Strophiurichthys robustus,* 23cm.
6. Weeping Toado, *Torquigener pleurogramma,* 13.5cm.
7. Spiny-backed Cow-fish, *Lactoria fornasini,* 13cm.
8. Striped Toby, *Canthigaster coronata,* 13cm.
9. Common Toado, *Torquigener hamiltoni,* 9cm.

PLATE 135

1. Long-spined Porcupine-fish, *Diodon holocanthus,* 30cm.
2. Myers' Porcupine-fish, *Dicotylichthys myersi,* 36cm.
3. Oblong Sunfish, *Ranzania laevis,* about 43cm.
4. Ocean Sunfish, *Mola mola,* 3m.
5. Bridled Cling-fish, *Lepadichthys frentus,* 5cm.
6. Banded Frogfish, *Halophryne diemensis,* 11cm.
7. Banded Frogfish, dorsal view of head.
8. Frogfish, *Batrachomoeus dubius,* 10cm.

PLATE 136

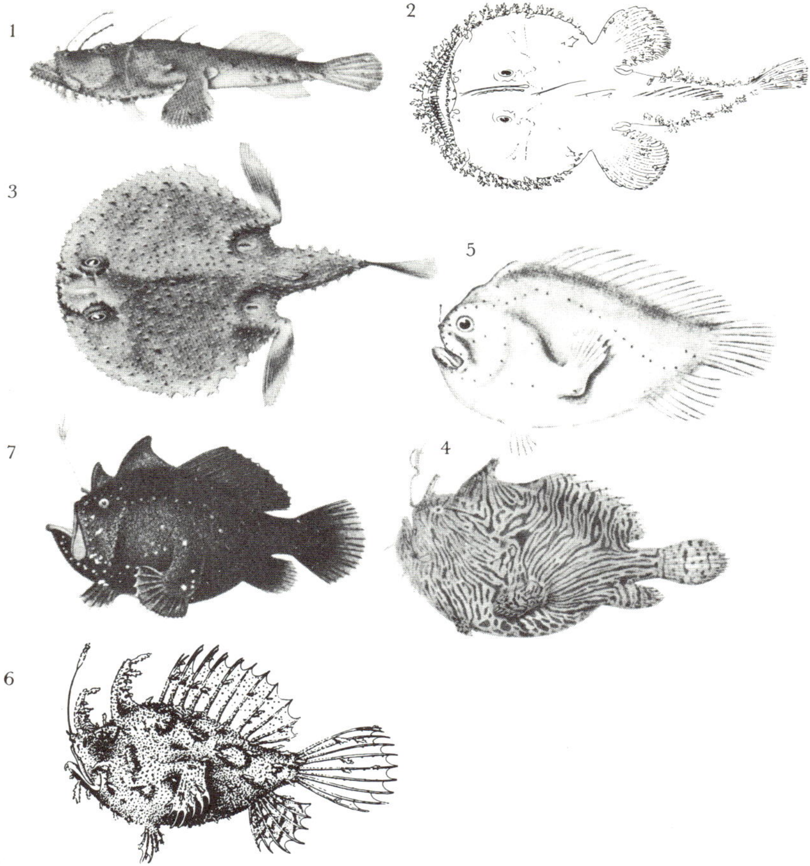

1. Broad-headed Fishing-frog, *Lophiomus setigerus,* 18.5cm, lateral view.
2. Broad-headed Fishing-frog, dorsal view.
3. Spiny Sea-bat, *Halieutaea brevicauda,* 10cm.
4. Striped Angler, *Antennarius tridens.*
5. Smooth Angler, *Histiophryne bougainvilli,* 3.5cm.
6. Harlequin Angler, *Tathicarpus muscosus,* 8cm.
7. Black Angler, *Antennarius moluccensis,* 10cm.